USES
OF
INFINITY

USES OF INFINITY

LEO ZIPPIN

Dover Publications, Inc.
Mineola, New York

Published in Canada by General Publishing Company, Ltd., 30 Lesmill Road, Don Mills, Toronto, Ontario.

Bibliographical Note

This Dover edition of *Uses of Infinity,* first published in 2000, is an unabridged republication of the book originally published under the same title in 1962 by Random House, Inc., New York.

Library of Congress Cataloging-in-Publication Data

Zippin, Leo.
    Uses of infinity / Leo Zippin.
        p. cm.
    Originally published: New York : Random House, 1962.
    Includes bibliographical references.
    ISBN 0-486-41178-8 (pbk.)
    1. Processes, Infinite. 2. Infinite. I. Title.

QA295 .Z5 2000
515'.24—dc21

                                                                99-059404

Manufactured in the United States of America
Dover Publications, Inc., 31 East 2nd Street, Mineola, N.Y. 11501

Contents

USES OF INFINITY

Preface

Most of this book is designed so as to make little demand on the reader's technical competence in mathematics; he may be a high school student beginning his mathematics now or one who has put away and forgotten much of what he once knew. On the other hand, the book is mathematical except for the first chapter—that is to say, it is a carefully reasoned presentation of somewhat abstract ideas. The reader who finds the material interesting must be prepared, therefore, to work for it a little, usually by thinking things through for himself now and then and occasionally by doing some of the problems listed. Solutions to some of these are given at the end of the book. But it will not pay the reader to stop too long at any one place; many of the ideas are repeated later on, and he may find that a second view of them leads to understanding where a first view was baffling. This style of presentation is imposed upon an author by the nature of mathematics. It is not possible to say at once all of the key remarks which explain a mathematical idea.

Many a reader is perhaps wondering whether it is possible for fellow human beings to communicate upon a topic as remote-sounding as "uses of infinity"; but, as we shall see, any two people who know the whole numbers,

$$1, \quad 2, \quad 3, \quad 4, \quad 5, \quad \cdots,$$

can talk to each other about "infinities" and have a great deal to say.

I have written this book from a point of view voiced in a remark by David Hilbert when he defined mathematics as "the science of infinity". An interesting theorem of mathematics differs from interesting results in other fields because over and above the surprise and beauty of what it says, it has "an aspect of eternity"; it is always part of an infinite chain of results. The following illustrates what I mean: the fact that $1 + 3 + 5 + 7 + 9$, the sum of the first five odd integers, is equal to 5 times 5 is an interesting oddity; but the

3

theorem that *for all* n the sum of the first n odd integers is n^2 is mathematics.

I hope that the reader will believe me when I say that professional mathematicians do not profess to understand better than anybody else what, from a philosophical point of view, may be called "the meaning of infinity." This is proved, I think, by the fact that most mathematicians do not talk about this kind of question, and that those who do do not agree.

Finally, I wish to express my especial thanks to Mrs. Henrietta Mazen, a teacher of mathematics at the Bronx High School of Science, who selected and edited the material in this monograph from a larger body of material that I had prepared. The reader who enjoys this book should know that in this way a considerable role was played in it by Mrs. Mazen. I am also indebted to Miss Arlys Stritzel who supplied most of the solutions to the problems posed in the book.

CHAPTER ONE

Popular and Mathematical Infinities

One who has not worked in the mathematical sciences is likely to doubt that there is any use of infinity if "to use" something means to acquire some form of control over it. But the use of infinity in precisely this sense constitutes the profession of mathematicians.

Other professions also use infinity somewhat. The architect and the engineer have their tables of trigonometric functions, logarithms, and the like. But they do not need to remember that these are calculated from a large number of terms of certain appropriate infinite series. They draw freely from an inexhaustible reservoir of mathematical curves and surfaces, but they do not need to be conscious of infinities. The philosopher and theologian are conscious of infinity, but from the mathematician's view they do not use it so much as admire it.

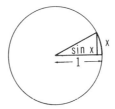

Figure 1.1. $\sin x = x - \dfrac{x^3}{6} + \dfrac{x^5}{120} - \dfrac{x^7}{5040} + \cdots$

The mathematician also admires infinity; the great David Hilbert† said of it that in all ages this thought has stirred man's imagination most profoundly, and he described the work of G. Cantor‡ as introducing man to the Paradise of the Infinite. But the mathematician also *uses* infinities and, as the next two chapters will show, he is the world's greatest collector of infinities—infinitely many arrays of infinities of all types and magnitudes. They are his raw materials and also his tools.

Before turning to mathematics, let us spend a moment with some examples of the popular "everyday" infinities which, as we shall see, are not too distantly related to the mathematical infinities. We begin with a folk-saying: "There are always two possibilities." Let us here call them "Zero" and "One". Since each choice made brings two new alternatives, this suggests the picture of infinity shown in Figure 1.2.

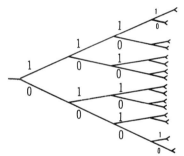

Figure 1.2. There are always two possibilities

Figure 1.3. An imitation of infinity

† David Hilbert (1862–1943) was one of the leading mathematicians of the twentieth century.

‡ G. Cantor (1845–1918) created set theory.

Next, many people can recall a certain box of baking soda on which, in their youths, they saw their first picture of infinity. This box had on it a picture of a box on which there was a picture of the same box showing another, and so on. Figure 1.3 conveys the impression this gave.

Next, there are intimations of infinity designed especially for children. Japanese artisans make wooden dolls that open up and contain a similar doll, inside of which there is another doll and so on through a sequence of five or six.

Also, poets have employed the word in ways that are not far removed from mathematical aspects of infinity. Juliet's line about her love for Romeo, "the more I give to thee, the more I have", exaggerates a characteristic property of cardinal infinities; to Blake's image, "To hold infinity in the palm of your hand and eternity in an hour", corresponds the mathematical fact that a segment as short as the palm's "life-line" has as many points on it as the infinitely long line. In *Anthony and Cleopatra*, General Enobarbus describes Cleopatra:"Time cannot wither nor custom stale her infinite variety." Hugo describes Shakespeare: "Genius is a promontory jutting into the infinite."

In a lighter vein, Rostand's Cyrano de Bergerac, among other fanciful schemes for going to the moon, uses a jocular version of mathematical induction: "I stand on a platform holding a strong magnet which I hurl upwards. The platform follows. I catch the magnet and hurl it up again, the platform following, and repeating this in stages, I ascend to the moon."

Very instructive are the metaphysical arguments of Zeno† directed to the conclusion that physical motion is impossible. He is quoted somewhat as follows: "Achilles cannot overtake a fleeing tortoise because in the interval of time that he takes to get to where the tortoise was, it can move away. But even if it should wait for him Achilles must first reach the half-way mark between them and he cannot do this unless he first reaches the half-way mark to that mark, and so on indefinitely. Against such an infinite conceptual regression he cannot even make a start, and so motion is impossible." Another pretty paradox of Zeno tries to show that it is impossible that space and time should *not* be infinitely divisible. Still another is usually quoted this way: "The moving arrow is at each instant at rest."

† A Greek philosopher of the Eleatic school who lived in the fifth century, B.C.

These paradoxes deal with an important application of mathematical infinities and deserve to be discussed here, the more so because the moving-arrow paradox is a neat formulation of the mathematical concept of motion. A *motion* is analogous to a time-table; more precisely, it is a *function* which associates a definite point in fictional "space" to each of certain moments in a fictional "time". From this point of view, the statement that the arrow is "at rest" at a given instant means that its position is defined; this gives the function. Functions defining a motion can be constructed like any other mathematical functions, that is to say by a suitable table of values, by a formula, or by a recursive description.

If Zeno's tortoise starts one foot ahead of Achilles and moves

$$\frac{1}{2} + \frac{1}{4} + \frac{1}{8} + \frac{1}{16} + \cdots \text{ feet}$$

(which is precisely one foot) in

$$\frac{1}{2} + \frac{1}{4} + \frac{1}{8} + \frac{1}{16} + \cdots \text{ seconds}$$

(which is precisely one second), while Achilles moves

$$1 + \frac{1}{2} + \frac{1}{4} + \frac{1}{8} + \frac{1}{16} + \cdots \text{ feet}$$

(which is precisely two feet) in

$$\frac{1}{2} + \frac{1}{4} + \frac{1}{8} + \frac{1}{16} + \frac{1}{32} + \cdots \text{ seconds}$$

(which is precisely one second), then the race is ended in one second.

Before the work of Eudoxus (350 B.C.) and of Archimedes (150 B.C.) these infinite series could not be understood. In the seventeenth century with the development of the calculus, the logic of infinite series had to be rediscovered. These *series* are probably not needed to "answer" Zeno, but they meet him very nicely on his own grounds.

$$\cdots, \frac{1}{9}, \frac{1}{8}, \frac{1}{7}, \frac{1}{6}, \frac{1}{5}, \frac{1}{4}, \frac{1}{3}, \frac{1}{2}, 1$$

Figure 1.4

The second of Zeno's paradoxes derives some of its interest from the fact that between every pair of points on a line there is another point. It follows that a segment contains an infinite *descending* sequence, that is to say an ordered sequence of points without a smallest term. The example in Figure 1.4 shows this; the fractions of the form $1/n$, arranged in the order of size, have no term that precedes all the others. No corresponding sequence exists in the set of whole numbers. It is likely that Zeno was entirely concerned with the problem which confronts applied mathematicians: that mathematics is an idealization of experience not necessarily "true to life". Nowhere is this more evident than in the simple fact that the geometric segment is infinitely divisible, but the matter in a material wire is not. Of course, this fact was not so convincingly established in Zeno's time as it is in our own. Nonetheless, in our time as in his, the mathematical segment serves as a model for many specific problems dealing with material bodies (vibrating strings, flexible beams, rigid bodies). Above all it serves as the model for the *time-continuum* and the one-dimensional *space-continuum*, and in this use it dominates our conception of the world about us.

$$\lim_{n \to \infty} \frac{n+4}{n} = \lim_{n \to \infty} \left(1 + \frac{4}{n}\right) = 1 + \lim_{n \to \infty} \frac{4}{n} = 1$$

Figure 1.5. A mathematical formulation of an everyday observation: Two friends—one four years older than the other—appear to approach the same age as the years pass

Finally let us look at a pair of everyday examples which have the flavor of mathematics. Anyone who has a friend several years older than himself notices how the passage of time thins out the difference between the two ages. We have here an example of a pair of variables whose difference is constant but whose ratio is not; in this example, the ratio approaches 1.

If	If
C denotes the cost	it costs \$8 to produce a sweater,
and	and
S denotes the selling price,	it is sold for \$12
then	then
$\dfrac{S-C}{C}$ is *per unit profit on cost,*	$\dfrac{12-8}{8} = \dfrac{1}{2}$ is *profit on cost,*
and	and
$\dfrac{S-C}{S}$ is *per unit profit on sales price.*	$\dfrac{12-8}{12} = \dfrac{1}{3}$ is *profit on sales price.*

Figure 1.6

The last example is drawn from commerce. When an object is bought for $1 and sold for $2, there is a 100 per cent profit if profit is figured on the purchase price; but there is a 50 per cent profit if this is figured on the selling price. An object can be sold at an arbitrarily large percentage profit on the purchase price, and it can be sold at 99 per cent or 99.99 per cent profit on the selling price, but it cannot, practically or theoretically, be sold at a 100 per cent profit on the selling price. The reader should persuade himself of this because it is a natural example of a situation in which one might be led to wonder about a "limit" which does not exist. Perhaps the easiest way to check on this particular problem is to try different selling prices.

Let us now turn to the uses of infinity in mathematics. It will help the reader to think of these as falling into four categories.

The first category is illustrated by the theorem of geometry: *If two sides of a triangle are equal then the base angles are equal* (Proposition 5, Book I of Euclid).

Proof. Given that $AC = BC$;† see Figure 1.7. Comparing the triangle ABC with itself, but reading it next as BAC, we find that $AC = BC$ and $\angle ACB = \angle BCA$. Therefore, by Proposition 4, Book I of Euclid, angle CAB must equal angle CBA.

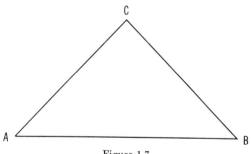

Figure 1.7

This proof that the base angles of an isosceles triangle are equal amounts to superimposing the triangle on itself so that A goes into B, B into A, and C into C.

The statement and proof depending on a proposition often referred to as "two sides and the included angle", or "s.a.s." (meaning "side-

† The statement $AC = BC$ means that the lengths of the line segments AC and BC are equal. In many books, the distance from a point P to a point Q is denoted by $\overline{PQ}$, but for reasons of typography it will simply be denoted by PQ in this book.

angle-side") make no mention of infinities. But the class of isosceles triangles (of all shapes and sizes) is an infinite class and the theorem holds for every one of them.

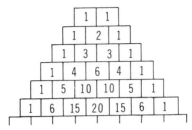

Figure 1.8. Pascal's Triangle. The nth row of this array gives the coefficients that occur in the expansion of $(a + b)^n$

The second category is illustrated by certain numbers, the "binomial coefficients", known to the Pythagoreans and to earlier civilizations, but associated with Pascal (1620) because of his use of mathematical induction in discussing them. The binomial coefficients, we recall, are the coefficients that occur when a binomial $a + b$ is multiplied by itself n times. For example, when $n = 3$, we have

$$(a + b)^3 = a^3 + 3a^2b + 3ab^2 + b^3,$$

and the binomial coefficients are 1, 3, 3, 1. Since n may be any positive whole number, we have an explicit infinity of cases, each of which involves a finite collection which we are invited to count.

The problem of finding a tangent line to a given curve at a point on the curve belongs to the third category. It is easy to see that this problem is associated with an infinite process, because tangency of a line to a curve can be determined by using arbitrarily small pieces of the curve and small segments of the line. It turns out that the mathematical process which solves this problem also solves the physical problem of defining *instantaneous velocity*; this is briefly speaking the number read on an automobile's speedometer. Velocity of motion and slope of a line are *limits* of *ratios*.

Figure 1.9. Tangency at a point is a local affair

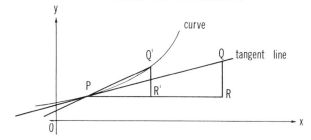

Figure 1.10. The *slope* of the tangent line (ratio QR/PR) is the *limit* of the slope of the chord (ratio $Q'R'/PR'$) as the point Q' approaches P along the curve

The fourth category belongs to abstract set theory and is concerned with infinite cardinal numbers. It is strikingly illustrated by the paradox that a circle can seem to have more points on it than the infinite line. Figure 1.11 shows this. Each point of the line is paired off against a point on the lower semicircle. Even if we match two points of the circle to one on the line there are still points (P and Q) left over.

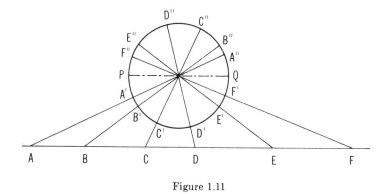

Figure 1.11

How infinity is handled in each of these cases is discussed throughout this book but can be described briefly as follows: To handle the first type, we use a single representative object; since it is in nowise special, it serves equally well for all. Problems of the second type are usually handled like the first but sometimes fall into a pattern that can be treated by mathematical induction, or the essentially equivalent "principle of infinite descent", also equivalent to the "principle of the first integer", as we shall see.

The third category includes a diversity of infinite processes associated with the notion of "limit". We shall study only one such process leading to a definition of *length* of certain figures; this in turn leads to a definition of the sum of an infinite series.

Although the summation of infinite series is closely related to the problem of defining (and calculating) the area bounded by a given closed curve, we shall not deal with area or with the earlier mentioned problem of tangency. We refer the reader to a good calculus book (see also Courant and Robbins, *What is Mathematics?* New York: Oxford University Press, 1941).

Problems in the fourth category called for the invention of a new type of mathematical reasoning and waited for the genius of Cantor. The arguments which Cantor employed are straightforward generalizations of the ordinary processes of counting, but he used them boldly—applying them directly to infinite sets. Perhaps his greatest single discovery was the fact that there are different cardinal infinities and that, in particular, the set of points on a line-segment is an incomparably "richer" infinity than the set of all whole numbers. This and other aspects of his work will be touched on in Chapters 2 and 6.

The book will close with some indications of the way in which this last aspect of infinity has been incorporated into the main mathematical stream of uses of infinity.

CHAPTER TWO

From Natural Numbers to $\sqrt{2}$

The main body of this chapter consists of a parade of short sections each of which concerns itself with some particular infinite set, infinite process, or point of view or technique for controlling infinity. We have tried to treat each of these separately, as much as possible; however, the first section begins with three infinities!

2.1 Natural Numbers

At the head of the parade of infinities, set off from the others, come the ordinary numbers. We are not going to be able to construct infinite sets, or prove anything significant about them, unless somehow we start with at least one infinite set, already constructed for us. Such a set exists, namely the ordinary numbers: 1, 2, 3, 4, 5, 6, $\cdots$ and so on. Concerning these, often called *natural numbers*, we are free to assume the following facts:

a) Each natural number has an immediate successor, so that the procession continues without end.

b) There is no repetition; each number is different from all the preceding numbers.

c) Every whole number can be reached in a *finite* number of steps by starting at 1 and counting up, one at a time, through the line of successors.

14

2.2 Discussion of Sequences

Since all numbers in the set of natural numbers cannot be written in a finite time we use "$\cdots$", that is, suspense dots or iteration dots, usually just three of them; they correspond to the words "and so on" or "et cetera", or "and so forth". They are used in mathematics following a short sequence of terms to mean that the indicated set is non-terminating and that the sequence is constructed according to a transparent scheme. The following examples will show what is meant; the reader is invited to put in the next few terms in each case.

1, 4, 7, 10, 13, $\cdots$,
5, 3, 5, 3, 5, 3, $\cdots$,
5, 3, 7, 4. 10, 6, 14, $\cdots$,
1, 1, 2, 3, 5, 8, 13, 21, 34, $\cdots$,
3, 6, 9, 12, 15, $\cdots$,
3, 9, 27, 81, 243, $\cdots$,
60, 3600, 216000, 12960000, $\cdots$,
2, 3, 5, 7, 12, 17, 29, 41, 70, 99, $\cdots$,
2, $\sqrt{2}$, $2+\sqrt{2}$, $\sqrt{2+\sqrt{2}}$, $2+\sqrt{2+\sqrt{2}}$, $\cdots$,
1, 6, 12, 20, 30, 42, $\cdots$,
1, 2, 6, 24, 120, 720, 5040, 40320, $\cdots$.

Our discussion of the use of iteration dots amounts to a definition of the idea of a sequence. This concept is so important to everything that follows that it is worth the reader's while to have it defined explicitly.

By definition all sequences are derived from the basic sequence

(1)   1, 2, 3, 4, $\cdots$

by replacing each number of (1) by some object. Thus, for example

i) a_1, a_2, a_3, $\cdots$

is a sequence of distinct letters,

ii) a, b, c, a, b, c, $\cdots$

is a periodic sequence of letters,

iii)   1, 1, 1, $\cdots$

is a sequence of 1's, and

iv) $\sqrt{6}$, $\sqrt{24}$, $\sqrt{60}$, $\sqrt{120}$, $\sqrt{210}$, $\cdots$

is a sequence of positive square roots.

Since a sequence is infinite we cannot be sure that we know what it is unless we have a rule which tells us how to replace each number in (1) by the appropriate object. In cases i), ii), and iii) the rule is quite well shown by the first three terms and the use of the iteration dots. In case iv), the reader will find that the terms can be calculated from this rule: The general term is the square root of the product of three consecutive integers the first of which is equal to the number of the term itself. Thus the fifth term of iv) is $\sqrt{5 \cdot 6 \cdot 7}$. The rule can also be expressed in a convenient formula. Let n designate the number of the term we seek and let us call this term a_n (read it: "a sub n"). Then

$$a_n = \sqrt{n(n + 1)(n + 2)}.$$

However, this is by no means an obvious formula and it is possible to get the first five terms by the use of many other formulas which would give different later terms. Therefore, in order really to know what series is meant in iv), one requires the explicit rule above. Whether this rule is expressed in words or by a nice formula is not important.

We can summarize our discussion symbolically by writing

$$a_n = f(n)$$

where a_n denotes the nth term of a sequence, that is, the object which replaces the number n in (1), and the symbol $f(n)$ denotes the rule for finding a_n, whether it is written as a formula or in words.

Thus in cases i) to iii):

i) $\qquad\qquad f(n) = a_n, \qquad\qquad n = 1, 2, 3, \cdots,$

ii) $\qquad\qquad f(n) = \begin{cases} a, & \text{for } n = 1, 4, 7, \cdots \\ b, & \text{for } n = 2, 5, 8, \cdots \\ c, & \text{for } n = 3, 6, 9, \cdots, \end{cases}$

iii) $\qquad\qquad f(n) = 1, \qquad\qquad n = 1, 2, 3, \cdots.$

There are instances when such dots have a different meaning. For example, we may write "the ten digits used in our decimal system are 0, 1, 2, $\cdots$, 9". Here the dots do not indicate an infinite set; they merely represent an abbreviation for the numbers 3, 4, 5, 6, 7, 8. Frequently mathematicians use dots to mean that a sequence is non-terminating, but is constructed according to some scheme the details of which are not important at the moment. Thus, in this sense,

$$\sqrt{2} = 1.414 \cdots \qquad \text{or} \qquad \pi = 3.14159 \cdots$$

means that $\sqrt{2}$ and π (pronounced "pie") are not finite decimals, and that they begin as shown. The reader will be able to tell the meaning of the dots from the context.

Problems

Comment on the following uses of "$\cdots$":

2.1. solos, duets, trios, quartets, $\cdots$.

2.2. singles, doubles, triples, home runs, $\cdots$.

2.3. twins, triplets, quadruplets, $\cdots$.

2.4. Monday, Tuesday, Wednesday, Thursday, Friday, $\cdots$.

2.5. M, T, W, T, F, S, S, M, T, W, $\cdots$.

2.3 Cardinals and Ordinals

The first member of our parade, the set of natural numbers, is connected with the next two infinite sets in the procession. The three infinite sets are:

a) The totality of whole numbers:
one, two, three, four, five, six, $\cdots$;
b) The collection of "rules" binding each to its successor:
one and one is two,
two and one is three,
three and one is four,
$\cdots$;
c) The aggregate of ordinal numbers:
first, second, third, fourth, fifth, sixth, $\cdots$.

The set a) being given to us, b) is a record of observations which we make on some and believe about all of the members of a). Because the rules are so transparently repetitive, they suggest that we can survey all of a) in our minds even though we can only exhibit a few of its members to our eyes. By making us conscious of the natural ordering of the whole numbers, the set c) hints at how these can be brought under our control, to some extent.

2.4 Arithmetic Sequences

The *augmented natural numbers* 0, 1, 2, 3, 4, $\cdots$ are obtained by putting zero in front of the sequence of ordinary whole numbers. We shall also call them *non-negative integers*, or simply *integers*.

If we begin with zero and write every second number thereafter, we get the sequence of the *even* integers:

$$0, \quad 2, \quad 4, \quad 6, \quad 8, \cdots \qquad \text{EVEN};$$

and if we begin with 1 and write every second number thereafter, we get the *odd* integers:

$$1, \quad 3, \quad 5, \quad 7, \quad 9, \cdots \qquad \text{ODD}.$$

A little thought will convince us that every natural number belongs to one or the other (not both) of these two sets: Every number is either even or odd. It will be useful to us to say this a little more elaborately: Every integer is of the form two-times-some-integer *or* it is of the form one-more-than-two-times-some-integer. Using convenient letters in place of words, we write: Every integer is of the form

$$n = 2q \qquad \text{or} \qquad n = 1 + 2q$$

where q is an integer.

Similarly, counting by threes, we find that every integer is of the form

$$n = 3q \qquad \text{or} \qquad n = 1 + 3q \qquad \text{or} \qquad n = 2 + 3q$$

where q is an integer.

Problem

2.6. This is equivalent to saying that every integer is in one of three non-overlapping infinite sets (and only one of them); show what these sets are.

In general, if B denotes any non-zero integer we can separate all integers into B distinct classes:

$$
\begin{array}{llll}
0, & B, & 2B, & 3B, \cdots, \\
1, & 1 + B, & 1 + 2B, & 1 + 3B, \cdots, \\
2, & 2 + B, & 2 + 2B, & 2 + 3B, \cdots, \\
\cdot & \cdot & \cdot & \cdot \quad \cdot \\
\cdot & \cdot & \cdot & \cdot \quad \cdot \\
B - 1, & (B - 1) + B, & (B - 1) + 2B, & (B - 1) + 3B, \cdots;
\end{array}
$$

the last line is less awkwardly written as

$$B - 1, \qquad 2B - 1, \qquad 3B - 1, \cdots.$$

Thus if B denotes 5, say, we get five $(=B)$ classes:

$$0, \quad 5, \quad 10, \quad 15, \cdots,$$
$$1, \quad 6, \quad 11, \quad 16, \cdots,$$
$$2, \quad 7, \quad 12, \quad 17, \cdots,$$
$$3, \quad 8, \quad 13, \quad 18, \cdots,$$
$$4, \quad 9, \quad 14, \quad 19, \cdots.$$

In some applications, these different classes are called *residue classes modulo B*; in the operation of division by B, they correspond to the remainders.

We can express the content of the above table as follows: Every integer n is of the form

$$n = qB + r$$

where q and r are integers; q is the multiplier (quotient, if we are thinking of division), r is the residue (remainder) and has one of the values in the finite set of values $0, \quad 1, \quad 2, \quad 3, \cdots, B - 1$.

This is the most important formula in arithmetic; in case $r = 0$, B is called a factor of n. (The value of q, as far as this definition goes, is not important.)

In the case $B = 10$, the residues correspond precisely to the digits $0, \quad 1, \quad 2, \quad 3, \cdots, \quad 9$.

Problem

2.7. Prove that n and $n + 1$ do not have a common factor, i.e., if an integer $q \neq 0, 1$ divides the integer n it does not divide $n + 1$.

2.5 Positional Notation

We now meet *positional notation*, our way of writing augmented numbers. This familiar scheme, on the base 10 (Hindu-Arabic notation), is illustrated by the development:

> nineteen hundred and sixty
> one thousand, nine hundreds, six tens, no units
> one, nine, six, zero
> 1, 9, 6, 0
> 1960.

It has these excellent features (not specific to the base ten):

    a) units of different sizes convenient for all uses,
    b) simple rules connecting consecutive units,

c) economy of expression with rapid growth of numbers represented, and

d) fixed set of digits.

Since only the digits and their relative positions figure in the writing
down of numbers, it follows that all of the operations of arithmetic
must be describable in terms of the digits, if proper account is taken
of position. The positional system, with base 60, was used already in
ancient Babylon. The Hindu-Arabic decimal system was made
popular in Europe in the thirteenth century principally by Leonardo
of Pisa, also called Fibonacci,† who learned it in his business travels
in Africa. He wrote a textbook in mathematics: *Il Liber Abaci*,
published in 1202; a second edition in 1228 survived for many centuries.

The formulation of arithmetic techniques in symbols is the beginning of algebra. That multiplication distributes over the summands,
as in the model below, is expressed by the equation:

$$a \cdot (b + c + d) = a \cdot b + a \cdot c + a \cdot d,$$

e.g., $4321 \times 567 = 4321 \times (500 + 60 + 7)$

$$= 4321 \times 500 + 4321 \times 60 + 4321 \times 7.$$

To use positional notation well, one must know the multiplication
table—at least up to "ten times ten". Although all of us know this
table, not many of us have looked at it as closely as we shall now.
The table is shown in Figure 2.1.

The rulings call attention to the fact that this table is a nest of
multiplication tables of increasing size, a one by one table, two by two
table, three by three, etc.

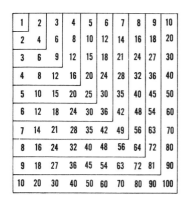

Figure 2.1

Figure 2.2

† Fibonacci means "son of good fortune".

The boomerang figures, or corner-frames in Figure 2.2, called *gnomons* by the Greek geometers, are also important for the examples which follow.

It was noticed, probably before 500 B.C., that the sum of the integers in any lower gnomon-figure is a perfect cube. Thus:

$$1 = 1^3, \qquad 2 + 4 + 2 = 8 = 2^3,$$
$$3 + 6 + 9 + 6 + 3 = 27 = 3^3, \quad \cdots.$$

The sum of the integers in any square table, whether two by two or three by three, and so forth, is a perfect square. Thus:

$$1 = 1^2, \qquad 1 + 2 + 2 + 4 = 9 = 3^2,$$
$$1 + 2 + 3 + 2 + 4 + 6 + 3 + 6 + 9 = 36 = 6^2, \cdots.$$

The table is also a picture of the double distributive law, as Figure 2.3 shows.

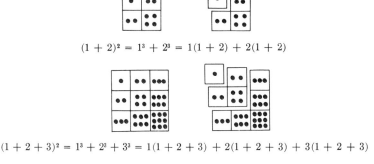

$$(1 + 2)^2 = 1^3 + 2^3 = 1(1 + 2) + 2(1 + 2)$$

$$(1 + 2 + 3)^2 = 1^3 + 2^3 + 3^3 = 1(1 + 2 + 3) + 2(1 + 2 + 3) + 3(1 + 2 + 3)$$

Figure 2.3

The proof of these facts for the given ten by ten table is a matter merely of verifying some simple sums and products. One naturally asks if the corresponding facts can be demonstrated for every k by k multiplication table where k is any positive whole number.

Problem

2.8. Verify that analogous facts are true for a 12 by 12 table or a 15 by 15 table. Can you see any inductive scheme, or general principle?

While positional notation has the advantages already mentioned, it also has one moderately troublesome feature, shown by the example

$$99999 + 1 = 100000;$$

the addition of one unit can make a very considerable change in many digits. All computing machines must take account of this, and probably everyone has witnessed the dials of a milometer turning together as the mileage on a car reaches 10,000.

In the case of decimals, the replacement of one number by another is common, $999.99 being thought a more attractive price than $1,000. Notice for later use that 1.000 000 and 0.999 999 differ only by .000 001, in spite of the substantial difference in their appearance. Thus

1.000	and	.999	differ by	.001;
1.000 000	and	.999 999	differ by	.000 001;
1.000 000 000	and	.999 999 999	differ by	.000 000 001.

The use of the decimals 0.1, 0.01, 0.001, $\cdots$ to express the tenth, hundredth, thousandth, $\cdots$ part of a unit allows us, in the denary positional notation, to express arbitrarily small numbers as well as arbitrarily large numbers. However, as the Babylonians may have been the first to discover (before 2100 B.C.), there are very useful numbers which cannot be expressed in this notation. The discovery was made in connection with "small" numbers, like $\frac{1}{3}$, $\frac{1}{6}$, $\frac{1}{7}$, $\cdots$, but it does not really depend on size, only on form; thus also $\frac{10}{3}$, $\frac{100}{3}$, $\cdots$ cannot be expressed in decimal notation. This inadequacy of the *finite* decimal system raises serious mathematical problems. One sees that the sequence 0.3, 0.33, 0.333, 0.3333, $\cdots$ represents numbers smaller than $\frac{1}{3}$ which approach $\frac{1}{3}$; clearly there is no terminating decimal which can represent $\frac{1}{3}$. However, from these same considerations, it follows that the symbol

$$0.33333 \cdots ,$$

a non-terminating decimal all of whose places are filled with 3's, must represent $\frac{1}{3}$, provided of course that this notation means anything at all.

The use of non-terminating decimals is the only way to achieve the *exact* representation of all numbers on the base 10. However, from the practical viewpoint of writing numbers or storing them in "memory tanks" on computing machines, it is necessary to sacrifice exactness and use the device of rounding. By carrying as many places as make sense for the problem in hand and specifying the range of error, one can be as accurate as necessary. The most common scheme of rounding entails an error not exceeding $\frac{1}{2}$ unit in the last place retained (5 units in the place following); all of this is shown in Figure 2.4.

If the product of two numbers is 1, then each of the two is called the *reciprocal* of the other. Thus since $\frac{1}{10}$ times 10 is 1, $\frac{1}{10}$ is the reciprocal of 10 and 10 is the reciprocal of $\frac{1}{10}$. In formulas, if the letter m stands for some number, and r for its reciprocal, one writes

$$r \cdot m = 1, \quad \text{and also} \quad r = \frac{1}{m}, \quad \text{and also} \quad m = \frac{1}{r}.$$

Integer	Common Fraction	To Three Decimal Places	To Five Decimal Places	To Seven Decimal Places	Reciprocal — Babylonian Scheme
1	$\frac{1}{1}$	1.000	1.00000	1.0000000	$\frac{60}{60}$
2	$\frac{1}{2}$	0.500	0.50000	0.5000000	$\frac{30}{60}$
3	$\frac{1}{3}$	0.333	0.33333	0.3333333	$\frac{20}{60}$
4	$\frac{1}{4}$	0.250	0.25000	0.2500000	$\frac{15}{60}$
5	$\frac{1}{5}$	0.200	0.20000	0.2000000	$\frac{12}{60}$
6	$\frac{1}{6}$	0.167	0.16667	0.1666667	$\frac{10}{60}$
7	$\frac{1}{7}$	0.143	0.14286	0.1428571	$\frac{8}{60} + \frac{34}{3600} + \frac{17}{216000} + \cdots$
8	$\frac{1}{8}$	0.125	0.12500	0.1250000	$\frac{7}{60} + \frac{30}{3600}$
9	$\frac{1}{9}$	0.111	0.11111	0.1111111	$\frac{6}{60} + \frac{40}{3600}$
10	$\frac{1}{10}$	0.100	0.10000	0.1000000	$\frac{6}{60}$

Figure 2.4. A table of reciprocals of the first ten integers, expressed to three, five, and seven decimal places

Since 0 (zero) plays a very special role in multiplication and $0 \cdot a = 0$ for all a, it follows that $0 \cdot a$ cannot also equal 1. Thus 0 has no reciprocal.

Problems

2.9. Show that $\frac{1}{7}$ gives rise to a non-terminating decimal by showing that the
expansion is periodic, i.e., repeating (the length of the period is six).
Similarly $\frac{1}{9}$, $\frac{1}{11}$, $\frac{1}{99}$ are periodic and necessarily non-terminating. Can
you prove that the terminating decimals correspond to fractions whose
denominators are products of a certain number of 2's and a certain number
of 5's $(2^m \cdot 5^n)$?

2.10. Can you identify the symbols $0.90909090909 \cdots$ and $0.090909090909 \cdots$?
What happens when you "add" these expressions as if they were numbers?
What do you expect to get? This non-terminating decimal form of the
number 1 is one of the minor problems that has to be cleared up before
one accepts non-terminating decimals as representing numbers.

2.11. Reduce the operation of division to the operation of multiplication by
means of a *table of reciprocals*; see Figure 2.4.

2.6 Infinities in Geometry

Euclidian geometry abounds in infinities. Some elementary theorems of plane geometry may be interpreted as theorems about
infinite sets.

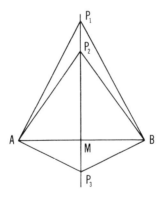

Figure 2.5

THEOREM 1. *Let AB denote a line segment. The locus of all points
in the plane equidistant from A and B is a line perpendicular to the
segment AB through its midpoint M.*

THEOREM 2. *If the three sides of one triangle taken in a definite order
are respectively equal to the sides of a second triangle taken in a definite
order, then each angle of the first triangle is equal to the corresponding
angle of the second triangle.*

In Theorem 1, the word "locus" (Latin for place) is just another word for "set": in the present case it refers to the set of points P in the plane such that the distance AP = distance PB. Here the letter P is used not just for one particular point, but to name any point at all that is as far from A as it is from B. The set of these points is an infinite set, and the theorem says that these points fill out a straight line. It also says that the new line is perpendicular to the old one, and the two intersect in the unique midpoint M of the segment AB. This new line is called the perpendicular bisector of the segment AB. You will recognize the fact that one way to find the midpoint M of AB is to construct this perpendicular bisector.

If you pick a point P such that $AP = PB$ and join that point P to the midpoint M of the segment AB you will get two triangles APM and BPM; you will see that

$$AP = BP, \qquad AM = BM, \qquad \text{and} \qquad PM = PM.$$

Now if you look at Theorem 2, you will see that it applies to this situation and tells you that $\angle AMP = \angle BMP$. Since the sum of these angles is a straight angle (180°), AMP and BMP are right angles (90°). This means that the line PM is perpendicular to the line AB, no matter which point P you may have chosen out of the infinite set for which $AP = PB$.

This does not conclude the proof of Theorem 1 (you can easily finish all the rest), but it accomplishes what we wanted to do. Namely, it proves something about every point of an infinite set without using an infinite process. In fact, it is not necessary, in the proof, that we be conscious of the fact that the set of points called P is an infinite set.

But we did need two important aids. First, we used the concept of *variable*. This is the name given to the letter P because of the way we used it. In our proof, the letters A and B, and the letter M, are not variables because of the way we used them. The letters A and B denote the same points throughout the proof. The letter M came about in the course of the proof. We may claim, if we like, that *we* named this point, but there is only one midpoint on AB, and M always means that midpoint.

But when we begin our proof, we suspect that there may be many different points, say P_1, P_2, P_3, and so on, such that

$$AP_1 = P_1B, \qquad AP_2 = P_2B, \qquad AP_3 = P_3B$$

and so on. When we use a single letter P to stand for *some* of them, we discover that what we are able to prove is also true for *every* one of them. And so we have really pulled off quite a sizable trick. The same form of words with the letter P standing now for the point P_1, now for the point P_2, now for P_3, and so on proves the same type of fact over and over again for *all* the points of our infinite point set. The letter P used this way is called a *geometric variable*, or just a *variable*.

If P is a variable point on the perpendicular bisector of the segment AB, then the length AP is also a variable, since it can refer to any one of many different lengths, and so also is the length PB. If we write

$$AP = PB$$

we have a kind of equation between variables, which the Greek mathematicians used quite regularly and which corresponds to our modern analytic geometry.

We needed variables to carry out our proof involving an infinite set, but this is not all we made use of. We also used the theorem above called Theorem 2. We said it applied to the triangles APM and BPM. But these triangles are variable, changing with different choices for P, and there is an infinite set of them. Fortunately, Theorem 2 covers this infinite set of possibilities.

The conclusion we can come to is that the theorems of geometry are indeed about infinite sets. But the proofs do not necessarily involve infinite processes, because the use of variables permits us to prove an infinite set of facts by a single argument.

2.7 Geometry Echoes Arithmetic

Euclid says, "If a line L and a segment $P'Q'$ are given, it is possible to lay off on L (in either direction) indefinitely often a segment PQ equal to $P'Q'$." The end point of the segment serves as the beginning point of the next segment.

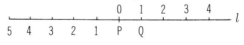

Figure 2.6

A sequence of points constructed in this way, going to the right (or left), is in an exact one-to-one correspondence with the set of all whole numbers. The whole numbers may be thought of as ordinals (i.e., as labelling the points in order, say from left to right) and they may be thought of as cardinals (i.e., as showing how many units PQ have been laid off on the line up to the point in question); see Figure 2.6. The two-way sequence corresponds, of course, to the positive and negative integers.

Some historians are of the opinion that the first books of Euclid's geometry were not meant to be the beginning of all geometry, but were meant to treat only those theorems which depended on ruler-and-compass constructions. Most of Euclid's postulates are in fact stated in terms of such constructions. From this point of view the postulate above merely asserts that, if one sets the compass for $P'Q'$ and chooses a point P as center on the line L, one can find the desired Q by swinging the compass; one can then find another point by choosing Q as center and swinging the compass, and thus go leap-frogging down the line.

Problems

2.12. Given a compass but no ruler, and two points P and Q, show that you can find the point R on the line through PQ such that $QR = PQ$, and show how to construct a sequence of points on a line, using only the compass.

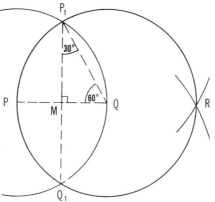

Figure 2.7

2.13. Suppose you have no compass, but a ruler only a little longer than PQ. Can you extend the line PQ indefinitely in each direction?

2.14. Explain how tunnelers run a road *straight* through a mountain.

2.8 Geometric Approach to Infinite Series

Euclid asserts that each segment may be bisected (meaning by ruler and compass) and that there results a pair of (equal) segments. Since each of these may be bisected, and each resulting segment, and so on, it follows that a segment contains infinitely many points. To be sure of this, all one has to do is make up some unambiguous scheme for a non-terminating process of bisections.

Figure 2.8 shows one particularly simple scheme: Each time one bisects the right-hand segment of the two new segments. Of course, during all of this construction one does not get to *B*, but that (Achilles' task) is not now the object. However, it is clear that, as one performs more and more bisections, the successive midpoints approach *B*; it is interesting that *every* systematic scheme for getting an infinite set by such repeated bisections has the property that successive midpoints actually approach some point of the segment.

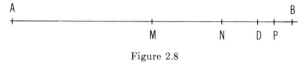

Figure 2.8

Problems

2.15. Continue the construction indicated in Figure 2.9 and decide what point of the segment the successive midpoints approach.

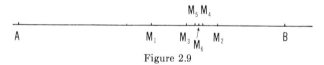

Figure 2.9

2.16. Using trisections, each time of the middle interval, convince yourself that the sequence of points constructed approaches the midpoint of the segment.

If the segment *AB* of Figure 2.8 represents an interval of length one, then the indicated bisections show rather convincingly that

$$\frac{1}{2} + \frac{1}{4} + \frac{1}{8} + \cdots = 1.$$

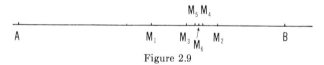

Figure 2.10

A picture of successive divisions into tenths in which the last tenth is always subdivided again (see Figure 2.10) shows that

$$.9 + .09 + .009 + .0009 + .00009 + \cdots = 1,$$

that is,

$$1 = 0.9999999 \cdots .$$

We shall discuss this more fully later.

Problems

Construct and consider:

2.17. $.3 + .03 + .003 + .0003 + \cdots$.

2.18. $\dfrac{2}{3} + \dfrac{2}{9} + \dfrac{2}{27} + \dfrac{2}{81} + \cdots$.

2.19. $\dfrac{1}{2} - \dfrac{1}{4} + \dfrac{1}{8} - \dfrac{1}{16} + \dfrac{1}{32} - \cdots$.

2.9 The Method of Exhaustion

Archimedes encountered the series

(2) $$\frac{1}{4} + \frac{1}{16} + \frac{1}{64} + \cdots = \frac{1}{3}$$

when he tried to find the area of a segment of a parabola. He perfected and applied to an infinite process the method of exhaustions devised by Eudoxus (450 B.C.). His proof of the equality (2) shows how certain things, obscure from one point of view, may seem plain from another. He said: "We divide a segment into four equal parts, we hold one for ourselves, give away two, and one remains. At this point three pieces have been handed out and we hold $\frac{1}{3}$ of the amount distributed. Next we take the remaining piece, divide it into four segments, keep one and give two away, so that again, one remains. After this step, it is again clear that we hold $\frac{1}{3}$ of all that has been distributed. Next we divide the remaining piece into four equal segments, hold one, give two away, and one remains. As we continue this process, we hold exactly $\frac{1}{3}$ of what has been handed out and a smaller and smaller segment remains to be distributed." This is the meaning of the equation above. In making the actual construction, it is best that the lengths which we want to add shall lie next to each other.

Figure 2.11

This method, which has in it all of the modern theory of limits, will be discussed more fully in Chapter 3. For the present, everyone will grant that it shows how the answer $\frac{1}{3}$ can be guessed; and also that, if any answer is right, $\frac{1}{3}$ is the one. Secondly, the argument shows that as one takes more and more terms of the series, the sum which one gets differs from $\frac{1}{3}$ less and less. Thirdly, this fact corresponds pre-cisely to the modern way of defining the *sum* of an infinite series *as a limit*, and so Archimedes' proof will seem to us entirely valid when we meet that definition.

Problem

2.20. Use this method to guess and prove your answers to:

a) $\dfrac{1}{5} + \dfrac{1}{25} + \dfrac{1}{125} + \cdots = ?$

b) $\dfrac{1}{8} + \dfrac{1}{64} + \dfrac{1}{512} + \cdots = ?$

c) $\dfrac{1}{10} + \dfrac{1}{100} + \dfrac{1}{1000} + \cdots = ?$

The same type of argument applies also to series like

$$\frac{2}{5} + \frac{4}{25} + \frac{8}{125} + \cdots .$$

Here a segment is divided into five parts, *two* are kept, two are worked on and one is given away; we hold $\frac{2}{3}$ of what is distributed and are led to an easy guess as to the answer.

Problem

2.21. Prove by the method of Archimedes, quite generally, for every pair of integers m and n with m *smaller* than $\frac{1}{2}n$, that

$$\frac{m}{n} + \left(\frac{m}{n}\right)^2 + \left(\frac{m}{n}\right)^3 + \cdots = \frac{m}{n - m} .$$

By substituting the letter r for the ratio m/n, we can write the left side of the formula in Problem 2.21 as

$$r + r^2 + r^3 + \cdots = ? .$$

Do you see how to get the right hand side in terms of r? If you can and if you have solved Problem 2.21, you will have a proof, by the method of Eudoxus-Archimedes, of the famous formula for the sum of all the terms of an infinite geometric progression whose ratio r is rational and smaller than $\frac{1}{2}$. However, there are many proofs of this formula valid for any r numerically less than 1.

Finally, convert your answer to

$$1 + r + r^2 + \cdots = \frac{1}{1 - r},$$

and observe that

$$a + ar + ar^2 + \cdots = \frac{a}{1 - r}.$$

Can $a = 0$ in this formula? Can $r = 1$ or can r exceed 1 in the original method, i.e., can $m = n$ or $m > n$? (The sign "$>$" is read "greater than".)

2.10 The Square Root of Two

The formula

$$\sqrt{2} + (\sqrt{2})^2 + (\sqrt{2})^3 + \cdots = \frac{\sqrt{2}}{1 - \sqrt{2}}$$

is absurd, as a moment's thought will show. The right hand side is negative; the left side, if it is anything at all, is "infinite". Although it has the correct form, namely,

$$r + r^2 + r^3 + \cdots = \frac{r}{1 - r}$$

it violates the stipulation that r be smaller than 1. On the other hand, $\sqrt{2}/2$ is less than 1 and the formula

$$\frac{\sqrt{2}}{2} + \left(\frac{\sqrt{2}}{2}\right)^2 + \left(\frac{\sqrt{2}}{2}\right)^3 + \cdots = \frac{\dfrac{\sqrt{2}}{2}}{1 - \dfrac{\sqrt{2}}{2}},$$

which is equal to the positive number $\sqrt{2}/(2 - \sqrt{2})$, is valid. But the type of proof given in the preceding section does not prove it, because that method works only if the numbers in the progression are ratios of whole numbers.

We shall show first that there is such a number as $\sqrt{2}$ by proving that $\sqrt{2}$ represents the length of the diagonal of a unit square. We shall do this by proving the theorem of Pythagoras. Next we shall show that $\sqrt{2}$ is not the ratio of any pair of integers.

The following proof of the theorem of Pythagoras is of the "BE-HOLD" type (i.e., it is revealed to the beholder when he glances at a diagram; see Figure 2.12):

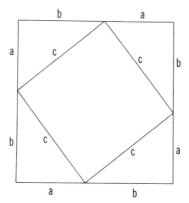

Figure 2.12. The inner square and the four triangles make up the big square

$$c^2 + 4\left(\frac{1}{2}\,ab\right) = (a + b)^2 = a^2 + b^2 + 4\left(\frac{1}{2}\,ab\right);$$

subtracting 4 ($\frac{1}{2}\,ab$) from the first and last member of this equality, we get

$$c^2 = a^2 + b^2.$$

Finally, if $a = b = 1$, we get

$$c^2 = 1 + 1, \qquad c = \sqrt{2};$$

and this is the length of the diagonal of a unit square.

The proof given below that $\sqrt{2}$ is not the ratio of two whole numbers is a so-called "indirect" proof, a reduction to absurdity (proof by contradiction, *reductio ad absurdum*).

If $\sqrt{2}$ could be represented as a ratio of two integers, say as p/q, then we could choose for p/q that fraction (among all equivalent fractions) which has the smallest possible denominator.

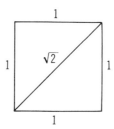

Figure 2.13

We know (see Figure 2.13) that 1 is smaller than $\sqrt{2}$, i.e.,

$$1 < \sqrt{2};$$

multiplying this inequality by $\sqrt{2}$, we also know that

$$\sqrt{2} < 2.$$

Now, if

$$\frac{p}{q} = \sqrt{2}, \quad \text{then} \quad p = \sqrt{2} \cdot q.$$

Hence p/q is larger than one but smaller than 2; that is, p is larger than q but smaller than $2q$, and so $p - q$ is positive and smaller than q.

Next, by squaring $p = \sqrt{2} \cdot q$, we have

$$p^2 = 2q^2,$$

and subtracting pq,

$$p^2 - pq = 2q^2 - pq \quad \text{or} \quad p(p - q) = q(2q - p)$$

so that

$$\frac{p}{q} = \frac{2q - p}{p - q}.$$

But the right side is a representation of p/q, with a smaller denominator than q, and this is a contradiction.

Problems

2.22. Discuss the same type of proof for the alleged irrationality of $\sqrt{3}$, $\sqrt{4}$, $\sqrt{5}$. Are these irrational? Proof? (Hint: In studying $p = \sqrt{5} \cdot q$, notice that $p - 2q$ is smaller than q since $\sqrt{5}$ is smaller than 3.)

2.23. Study $p = \sqrt{7} \cdot q$, noticing that $p - 2q$ is smaller than q.

2.24. Show that the irrationality of $\sqrt{8}$ follows easily from that of $\sqrt{2}$.

2.25. Study $p = \sqrt{n} \cdot q$ given that there is an integer k such that

$$k^2 < n < (k + 1)^2.$$

Show that if an integer is not the square of an integer, then it is also not the square of a fraction.

2.26. Show that $2\sqrt{s}$ is an integer only if $\sqrt{s}$ is an integer.

2.11 Computing the Square Root of Two

Let us conclude this chapter with two inductive constructions for approximations to $\sqrt{2}$. We need numbers like this in algebra in order to solve quadratic equations; we need $\sqrt{2}$ to solve

$$x^2 = 2.$$

The number may have been dreamed up by a person whose curiosity got the better of him as he looked at the sequence of squares

1, 4, 9, 16, 25, 36, 49, 64, 81, 100,

121, 144, 169, 196, 225, 256, $\cdots$

and noticed that there does not seem to be any square which is precisely twice as big as an earlier one. The pair 49 and 25 comes close, but it is a fact that no square integer is twice another square integer; that is,

$$m^2 = 2n^2$$

cannot be solved by any pair of integers. It *can* be solved by real numbers.

The problem of finding the *value* of this number, $\sqrt{2}$, has a very old history. Professor Neugebauer, one of the leading historians of mathematics, tells us that the Old-Babylonians gave the estimate (1; 24, 51, 10) in their sexagesimal notation, corresponding to 1.414213 $\cdots$ in our decimal notation, and that this estimate was used by the astronomer Ptolemy 2000 years later. He also indicates that a method still used for finding approximations to $\sqrt{2}$ (the first one which we are about to give) may have been the one that the Old-Babylonians used; the records suggest this, but are not conclusive.

The method is based on the following observation. Suppose that $a = \sqrt{2}$. Then

$$\frac{2}{a} = \frac{2}{\sqrt{2}} = \sqrt{2} = a$$

and therefore

$$a + \frac{2}{a} = 2a \quad \text{or} \quad \frac{1}{2}\left(a + \frac{2}{a}\right) = a.$$

If we knew the right value of $\sqrt{2}$ we could reproduce this value by taking one half of the sum: the value plus 2 divided by the value. Now for the method.

Step 1. Take a positive estimate, say 1 as first approximation.

Continuing instructions: Divide the integer 2 by the estimate obtained in the preceding step. Now take as a *new estimate* one half of the sum of this quotient and the estimate you just used.

The same general instruction in *algebraic notation* reads as follows: Let a denote the estimate to $\sqrt{2}$ obtained in the preceding step. Then the *new estimate* is

$$\frac{1}{2}\left(a + \frac{2}{a}\right).$$

We can also write the instructions as an inductive formula:

1st step $\qquad\qquad a_1 = 1,$

kth step $\qquad a_k = \frac{1}{2}\left(a_{k-1} + \frac{2}{a_{k-1}}\right), \qquad k = 2, 3, 4, \cdots.$

It is interesting to see what the first few approximations to $\sqrt{2}$ come out to be. We start with 1. Next we get $\frac{1}{2}(1 + 2) = \frac{3}{2}$. This estimate gives us next $\frac{1}{2}(\frac{3}{2} + \frac{4}{3}) = \frac{1}{2}(\frac{17}{6}) = \frac{17}{12}$. At the fourth step, $k = 4$ in the second formula, we get

$$\frac{1}{2}\left(\frac{17}{12} + \frac{24}{17}\right) = \frac{577}{408}.$$

This estimate is 1.414215 $\cdots$ and is reasonably close. A closer value would be 1.4142140.

It is clear that we get an infinite sequence of estimates in this way, but it is at least conceivable that after a certain number of steps we would find a fraction whose square is 2. If that could indeed happen, then the process would *continue* but it would give the same answer over and over. HOWEVER, that does not happen; it cannot happen because, as we saw in the preceding section, $\sqrt{2}$ is not a rational number (not the ratio of two integers). This fact was actually *proved* about 500 B.C., probably by Pythagoras himself. It is recorded that he considered the "irrationality" of $\sqrt{2}$ to be a most unwelcome fact, on philosophical grounds, because it showed that the world was not as simple and harmonious as he wanted it. He is supposed to have ordered the fact kept secret among the members of his own philosophical study-group, and there is a legend that at least one of his students was killed for spreading the bad news.

Pythagoras brought himself, and us, face to face with a serious question. Geometrical evidence says that there *is* such a number as the square root of 2, and it even says that this is a most important number *practically*, that is for carpentry and other technologies. But this number is certainly *not an integer*, and if it is also *not a ratio of integers*, then what kind of number is it, and how does one write it?

These are not hard questions to answer, nowadays, and this small book will answer them in time. When we want to *talk* about the $\sqrt{2}$, we simply give it any name by which other people will recognize it. For example, we could call it "that diagonal", or "the diagonal of a unit-square", or "the square root of 2" (we might add "the positive one", if we think there will be a misunderstanding, because there *is* a negative one); or today we would simply use the standard name for it, $\sqrt{2}$; (then we don't have to add "the positive one" because the notation includes that idea, since $\sqrt{2}$, by definition, stands for the *positive* root).

Of course, if we actually need to work with $\sqrt{2}$—for example if we are building something square—then the name itself is not going to be enough for us. But in that case we will know how accurate our work has to be and we will know therefore to how many places of decimals we want to have the $\sqrt{2}$. Then we will look it up in *tables of square roots*, and there are tables to five places and also tables to fifteen. Today a computing machine will get us a few thousand places in the course of an afternoon. If we have no machine and no tables, but plenty of paper, then the inductive construction above will get us as many places as we wish, if we go through enough steps.

If in this scheme one of the numbers a and $2/a$ is smaller than $\sqrt{2}$, then the other is larger than $\sqrt{2}$. For example, $a < \sqrt{2}$ implies

$$\frac{1}{a} > \frac{1}{\sqrt{2}} \quad \text{and} \quad \frac{2}{a} > \frac{2}{\sqrt{2}} = \sqrt{2}.$$

You can check quickly that $\frac{17}{12}$ is a little too big for $\sqrt{2}$, and that $\frac{24}{17}$ is too small. Now $\frac{17}{12}$ is about 1.417 and $\frac{24}{17}$ is about 1.413; if one of these numbers is too small and the other is too large then one half their sum (which is 1.415) must be correct to within two units in the last place. Now this may or may not be good enough for a particular job, but it shows the kind of thing the mathematician considers satisfactory: an *estimate* and then a *control* giving the range of possible error.

We present another way of finding approximations to $\sqrt{2}$ so closely connected to the decimal notation that it will help us to understand, later, what is meant by a *real number* ($\sqrt{2}$ is a real number; all rational numbers are real numbers also).

Step 1. Take the pair of consecutive integers 1 and 2. Multiply these by 10 (getting 10 and 20) and insert nine consecutive integers;

$$10, \quad 11, \quad 12, \quad 13, \quad 14, \quad 15, \quad 16, \quad 17, \quad 18, \quad 19, \quad 20,$$

is what you now have. Among these, there is a consecutive pair such that the square of the first has a leading digit 1, and the square of the second has a leading digit 2. *Find this pair of consecutive integers.* Either of these divided by 10 gives you an approximation to $\sqrt{2}$ (the first is a little too large and the second is a little too small).

Continuing step: Multiply the pair of consecutive integers obtained in the preceding step by 10 and insert the nine consecutive integers. Among the eleven integers there is a pair of consecutive ones such that the square of the first has a leading digit 1, and the square of the second has a leading digit 2. *Find this pair.* Either integer, divided by an appropriate power of 10, gives you an approximation to $\sqrt{2}$ (the first a little too small and the other a little too big).

To see how this works out, we calculate the squares of the first eleven integers up above:

$$100, \quad 121, \quad 144, \quad 169, \quad 196, \quad 225, \quad 256, \quad 289, \quad \cdots, \quad 400;$$

and now it is obvious that we want the square roots of 196 and 225, that is, the consecutive pair (14, 15).

The next stage gives 140, 150 and from these:

$$140, \quad 141, \quad 142, \quad 143, \quad 144, \quad 145, \quad 146, \quad 147, \quad 148, \quad 149, \quad 150.$$

By squaring we find that $(141)^2 = 19881$ and $(142)^2 = 20164$, so that the desired consecutive pair is (141, 142) and the associated approximations are 1.41 and 1.42. The reader will be wise if he carries this out for a few more steps. He will find that at each stage of the calculations, he can use the preceding calculations, and so systematize his work that it is almost pleasant. In addition, he will see that each successive step gives an improved answer; namely one more accurate place of decimals. Moreover, this method shows (the same fact is true for the earlier method but not as easy to see) that if n decimal places are wanted, they will always be obtained if n successive steps are calculated.

Problems

2.27. Prove the following rule: To multiply a number by 12.5, move the decimal point two units to the right, and then divide by 8.

2.28. The *root number* of a given number is obtained as follows: Add up all the digits; if the result is bigger than 9, add all its digits; continue until you get a result smaller than 10; this is the root number of the given number. Prove that a number and its root number belong to the same residue class modulo 3. Therefore, a number is divisible by 3 if and only if its root number is divisible by 3.

2.29. Using only 0's and 1's, (a) construct a decimal expansion with a very long period; (b) construct systematically a sequence of expansions, each with a period longer than that of the previous one; and (c) construct a non-terminating, non-periodic decimal.

CHAPTER THREE

From $\sqrt{2}$ to the Transfinite

3.1 Incommensurable Magnitudes

That $\sqrt{2}$ is irrational seems contrary to common sense, as the following will show: Suppose we are given two unequal lengths and are told, "Cut these into pieces of the same length, as many as you like, *but* they must all be of the same length!" It is surprising that this assignment is sometimes impossible to carry out. If the pieces are 27 and 33, say, or $27\sqrt{2}$ and $33\sqrt{2}$, the job is easy. But if the lengths are 1 and $\sqrt{2}$, the problem is insoluble.

One can get approximate solutions, but no exact solution. The irrationality of $\sqrt{2}$ shows that two given lengths are not necessarily *commensurable*; that is, it may happen that no unit of length exists such that each given length is a whole number of units. In order to compare two lengths, one measures them with the same ruler. This makes "incommensurability" of lengths sound like "incomparability" of lengths, and this is very disturbing. Indeed, it does imply the need for an infinite process.

The Greek mathematicians realized that the problem of calculating the perimeter and area of a circle also leads to an infinite process, since no replicas of a square, however small, can evenly fill out the area of a circle. The nature of the two shapes does not permit it; see Figure 3.1.

Thus mathematicians knew the difference between commensurability and comparability, but they were surprised to see this question come up in connection with straight line segments, and in connection also with the area of rectangles, as the next picture illustrates.

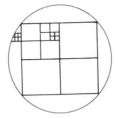

Figure 3.1

The rectangles in Figure 3.2 have two square feet of area. The first is 2′ long, 1′ wide; the second is $\frac{3}{2}′$ long, $\frac{4}{3}′$ wide; the third (the inner figure) is a square, $\sqrt{2}′$ on a side.

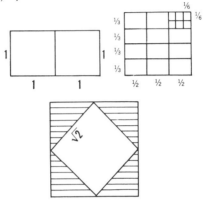

Figure 3.2

It is easy to see that the first contains two unit squares and the second contains 72 small squares (each $\frac{1}{6}$ by $\frac{1}{6}$). Also, the first can be divided into 72 ($\frac{1}{6}$ by $\frac{1}{6}$) squares. The third picture does not suggest any reason why there is *no* small square, replicas of which would evenly fill both the unshaded area and also the larger area. But this is what the irrationality of $\sqrt{2}$ entails.

Once the irrationality of $\sqrt{2}$ had been established, it was something the Greek mathematicians could get used to. But the embarrassing fact remained that since the Greeks had not originally conceived of this possibility, they had not allowed for it in their proofs. Many proofs, particularly in the theory of areas and of proportion, correct as far as they went, omitted what came to be called the "incommensurable case". (As recently as fifty years ago, high school text books in geometry featured careful treatments of the incommensurable case.) A very clear example of a theorem where the incommensurable case calls for a separate treatment is given in Chapter 6.

The corrected proofs called for a suitable theory of *limits*. This was supplied by Eudoxus (450 B.C.); he formulated a general principle now known as the principle of Eudoxus. A special form of it is the Axiom of Archimedes, which we now know to be too sweeping, but which was valid for all the geometrical configurations with which Eudoxus was familiar. Eudoxus' principle was this: If A and B are two given magnitudes of the same geometric kind (lengths, areas, or volumes), then there always exists some whole number, say m, such that m times A is larger than B.

He perfected a method of carrying out the limit process implied in his principle, which came to be known as the *method of exhaustions*. An example of this method was shown in Section 2.9.

It was Archimedes who, centuries later, used the work of Eudoxus to calculate the volumes and areas of a large variety of curved figures: the sphere and the circular cylinder; the parabolic cylinder and certain parabolic solids; the areas and lengths associated with certain spirals. The tradition of his accomplishments was passed on from generation to generation, and from country to country (leaving Europe about 500 A.D. and returning in later centuries, about 1400 A.D.). The rigorous thinking which lay behind it was somewhat lost sight of in the excitement of the generality of new algebraic techniques; it was rediscovered only in the nineteenth century.

What Eudoxus asserts is equivalent to this: Given two geometric objects (of the same kind) of magnitudes A and B, then for some integer q,

$$A = qB + R$$

where R is a magnitude of the same kind, smaller than B. By treating B as a variable and taking smaller and smaller units for B, one gets smaller and smaller remainders R, hence better and better approximations to the magnitude of A.

The fact that some geometric magnitudes had to be calculated by the use of infinite processes led to the converse question: How can one be sure that a given infinite process actually leads to a number? Eudoxus invented a suitable theory of limits, at the same time discovering the real number system. This was rediscovered by Dedekind in the 1870's.

Problem

3.1. Study the following theorem: Given triangle ABC and segment $B'C'$ parallel to the base BC. Then the ratios $AB':B'B$ and $AC':C'C$ are equal.

The dotted lines show how this can be proved in the special case

$$AB':B'B = 3:2.$$

The dotted angles are equal.

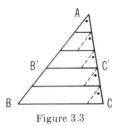

Figure 3.3

Generalize the proof to the case

$$AB':B'B = m:n$$

where m and n are positive integers. What happens if $m = \sqrt{2}$ and $n = 1$?

3.2 Geometric Constructions of Reciprocals and Other Number Relations

Every rectangle whose sides x and y have the product $xy = 1$ has an area of 1 square unit. If x is a given length, we can find y by a simple construction shown in Figure 3.4. First write:

$$y = \frac{1}{x}, \qquad \text{or better still} \qquad \frac{y}{1} = \frac{1}{x},$$

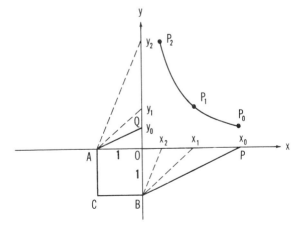

Figure 3.4

and interpret the last equation as an equality of the two ratios belonging to the similar triangles OPB, with $OP = x$ and $OB = 1$; and OQA, with $AO = 1$ and $OQ = y$. AQ is parallel to BP.

In this construction, $x = OP$ is given, the square $AOBC$ being fixed once and for all. Now one draws or merely imagines BP and constructs the parallel AQ, obtaining OQ as y. If one constructs such lengths y_0, y_1, y_2, $\cdots$ for many different given lengths x_0, x_1, x_2, $\cdots$ and uses the resulting pairs of numbers (x_k, y_k) as the coordinates of points P_k, one may connect them by a graph (when enough have been plotted to give one confidence in the resulting Figure 3.5).

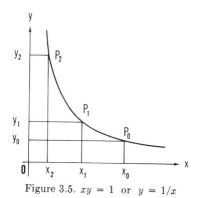

Figure 3.5. $xy = 1$ or $y = 1/x$

When used with a ruler, this figure becomes a table of reciprocals and may be used for determining the reciprocal $1/x$ for each number x.

Problems

3.2. Is there a rectangle of unit area, one of whose sides has length zero?

3.3. Analyze the process indicated by Figure 3.6 for finding the reciprocal $y = 1/x$ of a given number x.

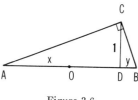

Figure 3.6

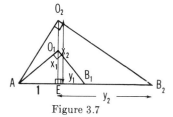

Figure 3.7

3.4. Work out the details of the construction of the parabola $y = x^2$, interpreted as

$$\frac{y}{x} = \frac{x}{1};$$

then construct the curve $y = x^2$ by graphing, using Figure 3.8 (with right angles at 0) for the plotting of points.

3.5. What is the bearing of this construction of the curve $y = x^2$ on the question of whether one can always find $\sqrt{x}$, given any length x. Relate this problem to the previous construction.

3.6. It is a fact that one cannot find cube roots, in general, by any finite number of constructions involving ruler and compass. But one can plot a great many "cubic numbers", and draw a curve $y = x^3$. How can this curve be used to find cube roots approximately?

3.7. Construct $y = x^2$ and read off

$$\sqrt{2}, \qquad \sqrt{2 + \sqrt{2}}, \qquad \sqrt{2 + \sqrt{2 + \sqrt{2}}}.$$

Use the curve and a ruler to solve $x^2 + x - 1 = 0$, in the form $x^2 = 1 - x$, i.e., $x^2 = y = y' = 1 - x$.

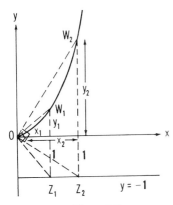

Figure 3.8

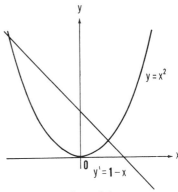

Figure 3.9

3.3 A Whirl of Quadratic Irrationals

In the sequence 1, $\sqrt{2}$, $\sqrt{3}$, $\sqrt{4}$, $\sqrt{5}$, $\sqrt{6}$, $\cdots$ some are obviously whole numbers and therefore rational, e.g.,

$$\sqrt{4} = 2 = \frac{2}{1}.$$

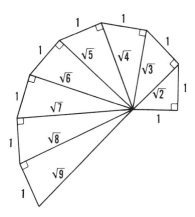

Figure 3.10. Whirl of Irrationals

It is not difficult to prove that only these, the "square numbers", are rational. Figure 3.10 indicates how all of them can be constructed, inductively, beginning with 1 and building right-angled triangles.

Problems

3.8. Carry out the construction until you feel able to guess what happens to the quantity $\sqrt{n+1} - \sqrt{n}$ as n gets larger and larger. Each term, of course, is bigger than the preceding term, and the difference between each term and its predecessor is always greater than zero.

3.9. Can you apply the identity $(a + b) \cdot (a - b) = a^2 - b^2$ to the problem above to help you prove your guess? Hint: If you use it correctly, the right hand side will be equal to 1.

3.10. Anticipate the behavior of $\sqrt{n}(\sqrt{2n+1} - \sqrt{2n})$ as n gets larger and larger.

3.4 Illustration of a Limit Point

The curve shown in Fig. 3.11 introduces the idea of a *limit* in a different way from any other example in this book. It was constructed

by Hippias, contemporary of Socrates (450 B.C.), and was used by him
to solve the problem of measuring the area of a circle. Its name,
"quadratrix", was derived from the fact that it "squared" the circle.

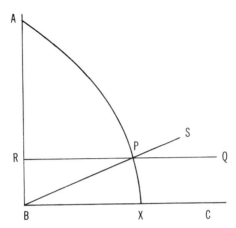

Figure 3.11. Quadratrix of Hippias

We shall use the curve here in a completely different context; we
shall examine the peculiar way in which one of the points of the
quadratrix is related to its other points. But first we shall describe
how this curve may be constructed.

The quadratrix is shown in Figure 3.11 as the arc AX. It is the
locus of points P found as follows: The right angle ABC and the
segment of length AB are given; then the point P is the point of
intersection of two straight lines, the "radial" line BS through B
and the horizontal line RQ parallel to BC, drawn in such a way
that the proportion

$$\angle CBS : \angle CBA = BR : BA$$

is satisfied. The angles must, of course, be measured in the same units
of angle, the lengths in the same units of length, so that each side of
the equation is a "pure" number (without units). It is convenient to
choose BA as the unit of length, and the right angle CBA, called
a *quarter-turn*, as the unit of angle. If we do this, the equation reads

number of quarter-turns in $\angle CBS$

$$= \text{number of units of length in } BR.$$

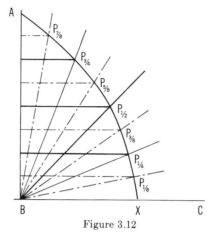

Figure 3.12

To construct this curve one first bisects the angle ABC and the line segment AB. This locates a point on the quadratrix which we shall call $P_{1/2}$ (read it "P sub one-half"). Next one bisects each of the two resulting segments along AB and also each of the two resulting angles $ABP_{1/2}$ and $P_{1/2}BC$ and locates two points, $P_{1/4}$ and $P_{3/4}$ (see Figure 3.12). Next one bisects each of the four new angles and the four new segments, and the correct association of horizontal and radial lines determines four new points, $P_{1/8}$, $P_{3/8}$, $P_{5/8}$, $P_{7/8}$ (listed counter-clockwise from the lowest). It is clear how one can find more and more points, as long as one wishes. At a certain stage in the actual execution of the construction, the curve more or less materializes. When we have plotted enough points, we can draw in the rest. The complete curve has now been constructed.

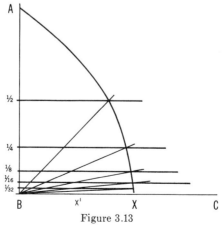

Figure 3.13

The point X on the base CB has an *exceptional* relation to the curve. It is on the radial line of zero quarter-turns and on the horizontal line of zero altitude. But these two lines whose intersection should have given us the point X *coincide!* Therefore the point X cannot be defined in the same way as the other points of the curve. One sees from the entire curve, once drawn, that it ought to be easy to locate the point X on CB; but the previous definition of the quadratrix does not apply to this point.

The same difficulty arises when one goes over to the analytic equation, as we shall see below. Meanwhile, Figure 3.13 shows how the point X can be calculated as a *limit* of points of the rest of the curve. This suggests that the point X on the quadratrix has to be *defined* as a limit of other points of the curve. This can actually be done and X can be found from this new definition.

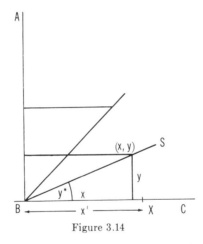

Figure 3.14

If we denote the coordinates of the points on the quadratrix by (x, y) and the angle CBS by y^* (the superscript* being read "quarter-turns"), we may use the definition of the tangent of y^* and write

$$\frac{y}{x} = \tan y^* \qquad \text{and} \qquad x = \frac{y}{\tan y^*};$$

See Figure 3.14.

Now the point X which interests us corresponds to $y = 0$; let the distance BX be called x'. The equation

$$x = \frac{y}{\tan y^*} \qquad\qquad \text{for} \quad y = y^* = 0,$$

takes the *indeterminate* form $x' = \frac{0}{0}$; clearly this tells us nothing about x'.

We can try to find x' as the following limit:

$$x' = \text{limit of } \frac{y}{\tan y^*} \text{ as } y \text{ approaches zero.}$$

To calculate this limit, we choose some sequence y_1, y_2, y_3, $\cdots$ of values of y approaching zero (in Figure 3.13 we used $\frac{1}{2}$, $\frac{1}{4}$, $\frac{1}{8}$, $\cdots$) and then study the sequence

$$\frac{y_1}{\tan y_1{}^*}, \qquad \frac{y_2}{\tan y_2{}^*}, \qquad \frac{y_3}{\tan y_3{}^*}, \qquad \cdots .$$

It turns out that, in the quarter-turn units of angle, this limit is $2/\pi$; in radian measure, the limit is 1. (The reason for the difference is that π radians are 2 quarter-turns.)

Problem

3.11. Figure 3.15 suggests the actual relative sizes of $\sin y$, y, and $\tan y$ when y is expressed in radians. Can you discover the relation of $y/\tan y$ to 1, for small values of y?

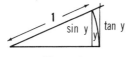

Figure 3.15

3.5 Limits

A sequence of numbers x_1, x_2, x_3, $\cdots$ has *the limit* x provided the sequence of numbers $(x_1 - x)$, $(x_2 - x)$, $(x_3 - x)$, $\cdots$ *approaches zero.*† For example, the sequence

a) $$3, \quad \frac{5}{2}, \quad \frac{7}{3}, \quad \frac{9}{4}, \quad \cdots$$

has the limit 2 because the sequence

$$3 - 2 = 1, \quad \frac{5}{2} - 2 = \frac{1}{2}, \quad \frac{7}{3} - 2 = \frac{1}{3}, \quad \frac{9}{4} - 2 = \frac{1}{4}, \quad \cdots$$

approaches zero. Similarly the sequence

b) $$\frac{1}{3}, \quad 3, \quad \frac{5}{7}, \quad \frac{7}{5}, \quad \frac{9}{11}, \quad \frac{11}{9}, \quad \cdots$$

† The expression "a sequence of numbers approaches zero" will be defined presently.

has the limit 1 because the sequence

$$\frac{-2}{3}, \quad 2 \quad \frac{-2}{7}, \quad \frac{2}{5}, \quad \frac{-2}{11}, \quad \frac{2}{9}, \quad \cdots$$

approaches zero. A more difficult example, requiring insight into the
nature of the sequence and also some computation, is the sequence

c) $\qquad \frac{3}{2}, \quad \frac{7}{5}, \quad \frac{17}{12}, \quad \frac{41}{29}, \quad \frac{99}{70}, \quad \cdots ;$

it has the limit $\sqrt{2}$. More difficult to understand, but requiring no
calculation of any kind, is the fact that the sequence

d) .1, .102, .102001, .1020010002, .102001000200001, $\cdots$

has a limit. This limit is defined by the sequence itself, and is otherwise unknown to us!

To complete the preceding definition, one needs a formal statement of what it means to say that a sequence approaches zero, or,
equivalently, that a sequence has zero as limit. Here one sees very
explicitly the use of infinity.

DEFINITION. A sequence of positive numbers

$$e_1, \quad e_2, \quad e_3, \quad \cdots$$

has the limit zero (approaches zero) if zero is the only non-negative
number which is smaller than infinitely many of the terms in the
sequence.

A general sequence approaches zero if it approaches zero when the
signs are all made positive.

This is more usually stated as follows:

EQUIVALENT DEFINITION. A sequence of numbers *has the limit zero*
provided that for every unit fraction $1/n$, that is

$$1, \quad \frac{1}{2}, \quad \frac{1}{3}, \quad \frac{1}{4}, \quad \cdots ,$$

there are at most a finite number of terms of the sequence which are
numerically larger than $1/n$ (the sign of the terms is not relevant).

Ultimately the usefulness of a theory of limits depends upon three
factors: In logical order these are 1) a precise definition of limit of a
sequence; ii) in each application, a proof of existence of the limit, and
iii) a method for calculating the limit; this frequently consists of some
scheme of successive approximations coupled with an estimate of

error. Of the preceding examples, a) and b) satisfy iii) so well that i) and ii) are superfluous, the idea of limit being almost self-evident in these cases. Example c) is a little more difficult; here one has the definition, knows $\sqrt{2}$, and is given a sequence of alleged approximations but has to work out some estimate of error. Without such estimates, one does not know what is meant by "approximation." The difficulty in example d) is of another kind. There the form of the approximation shows the nature of the error, but the question is one of *existence*: Is there a number which is being approximated? The next sections will show the meaning and usefulness of this question.

A few of the important theorems in a theory of limits, chosen because we shall need them very soon, read as follows:

THEOREM 3.1. *If* x_1, x_2, x_3, $\cdots$ *has the limit* L *then every subsequence has the same limit.*

For example, x_2, x_5, x_9, x_{14}, $\cdots$ has the limit L.

THEOREM 3.2. *If* x_1, x_2, x_3, $\cdots$ *has the limit* L, *then* $a + x_1$, $a + x_2$, $a + x_3$, $\cdots$ *has the limit* $a + L$.

For example, $-3 + x_1$, $-3 + x_2$, $-3 + x_3$, $\cdots$ has the limit $-3 + L$.

THEOREM 3.3. *If* x_1, x_2, x_3, $\cdots$ *has the limit* L, *then* kx_1, kx_2, kx_3, $\cdots$ *has the limit* kL.

For example, $2x_1$, $2x_2$, $2x_3$, $\cdots$ has the limit $2L$.

THEOREM 3.4. *If* x_1, x_2, x_3, $\cdots$ *has the limit* L, *then* x_1^2, x_2^2, x_3^2, $\cdots$ *has the limit* L^2:

Problem

3.12. Prove these theorems. More generally, show that x_1^n, x_2^n, x_3^n, $\cdots$ has the limit L^n for every integer n. [This is rather difficult, and the following course is suggested. Suppose that x_1, x_2, x_3, $\cdots$ is a sequence with a limit L, and that y_1, y_2, y_3, $\cdots$ is a sequence with a limit M. Prove, now, that the sequence of products x_1y_1, x_2y_2, x_3y_3, $\cdots$ has the limit LM, i.e. that the limit of the products is the product of the limits. This is also difficult; one must use the following type of technique:

$$LM - xy = LM - xM + xM - xy = (L - x) \cdot M + x(M - y).]$$

3.6 How the Fact that a Limit Exists May Help to Determine It

It is not obvious that the sequence of numbers

$$\sqrt{2}, \quad \sqrt{2 + \sqrt{2}}, \quad \sqrt{2 + \sqrt{2 + \sqrt{2}}}, \quad \cdots$$

has a limit, and the reader may wish to calculate a number of terms (one can use a well-drawn $y = x^2$ curve as a table of square roots and a ruler for measuring and transferring lengths) to anticipate what this limit may be.

Now, assuming merely that the sequence has a limit, let us call this limit x^* and look for it. If we square each term of the given sequence, we shall get a new one whose limit, according to Theorem 3.4, must be x^{*2}. But what we actually get is the sequence

$$2, \quad 2 + \sqrt{2}, \quad 2 + \sqrt{2 + \sqrt{2}}, \quad 2 + \sqrt{2 + \sqrt{2 + \sqrt{2}}}, \quad \cdots$$

each of whose terms (after the first) is 2 more than a term of the original sequence. Clearly the limit of the new sequence is $x^* + 2$. So now we know that

$$x^{*2} = x^* + 2,$$

and there are only two numbers for which this can be true: 2 and -1. So x^* is 2 *or* it is -1. Since all terms are positive, -1 is a ridiculous answer and it follows that $x^* = 2$.

3.7 Limits which Fail to Exist

Let us begin by making the foolish assumption that the sequence of numbers 1, 2, 3, 4, $\cdots$ has a limit, although we know that it does not, and let us call this "limit" x. If we double each element of the sequence we shall get a new sequence whose limit must be $2x$. But the new sequence is a subsequence of the original sequence and hence must have the original limit (see Theorem 3.1). Therefore, it follows that

$$x = 2x.$$

But the only number of this kind is zero. Therefore $x = 0$. We seem to have proved that the sequence of natural numbers approaches zero!

Of course, what we proved (by *reductio ad absurdum*) is that the given sequence has no limit.

Next let r be some number and look at the sequence

$$r, \quad r^2, \quad r^3, \quad \cdots .$$

Let us assume that it has a limit R. Since the sequence obtained by multiplying each term by r is a subsequence of the original, it follows that

$$R = R^2$$

and therefore $R = 0$ or $R = 1$. Both cases are possible. But there is another very important possibility, and that is that we are making a mistake. The sequence may not have a limit; in fact, if r is numerically larger than 1, no limit exists. Also, if $r = -1$ no limit exists.

Problem

3.13. (a) Show that, if $-1 < r < 1$, the limit of r, r^2, r^3, $\cdots$ is 0, and if $r = 1$, the limit is 1.

(b) In Section 2.9 we showed how to prove that $a/(1 - r)$ is the formula for the sum of the infinite geometric series $a + ar + ar^2 + \cdots$, whose ratio r is rational and numerically smaller than $\frac{1}{2}$. Now can you prove that this formula is valid when the common ratio r is irrational and numerically less than 1? Hint: Use the identity

$$1 - x^n = (1 - x)(1 + x + x^2 + \cdots + x^{n-1}).$$

3.8 The Real Number System and the Bolzano-Weierstrass Theorem

In the real number system, which will be described presently, the following fundamental theorem governs the subject of limits.

BOLZANO-WEIERSTRASS THEOREM. *If* x_1, x_2, x_3, $\cdots$ *is an increasing sequence of numbers, and if there exists a number* B *which is larger than all numbers of the sequence, then there is a number* L *such that* L *is the limit of the sequence.*

This theorem assures us, for example, that the sequence d)

$$.1, \quad .102, \quad .102001, \quad .1020010002, \quad \cdots$$

of Section 3.5 has a limit. For B we can choose any number larger than .2 or .11, or .103, etc. The theorem merely requires that we find some one number that is larger than all the numbers of the sequence. This condition rules out sequences like 1, 2, 3, 4, $\cdots$ (which, of course, do not have limits). The theorem does not tell us anything about the number L; for example, the numbers x_n may all be rational and L may be rational or irrational.

There are many equivalent ways in which mathematicians now set up the necessary axioms and definitions to construct the real number system. In some of these the proof of the Bolzano-Weierstrass principle is difficult, in others easy. In fact, there are so many details to take care of in setting up the real numbers, that if one fundamental theorem is made easy to prove, another one will be found difficult. Thus it turns out to be entirely correct to define the real number system as the smallest number system which contains all tne rational numbers and in which the Bolzano-Weierstrass theorem is true. It is now instantly clear that in this system we have got our theorem for free. Note that the Bolzano-Weierstrass theorem is false in the system consisting only of rational numbers.

A more satisfactory scheme from certain other points of view is the following: We agree to write all terminating decimals with an infinite string of zero's and then to define the real number system as the totality of all infinite decimal expansions; this includes the genuinely nonterminating ones. If we do this, we shall now have to prove the Bolzano-Weierstrass theorem, and we shall also have to show that we have a number system. That is, we shall have to show how to add infinite decimals, how to multiply them, and prove all the standard rules governing these operations. This calls for a great deal of work.

The Bolzano-Weierstrass theorem was not proved until 1865 when the calculus was over two hundred years old; one might wonder what mathematicians did about the existence of limits before that.

3.9 The Two Principal Cardinal Infinities

The set of integers and the set of all real numbers represent infinite cardinal numbers of different magnitudes, called *Aleph-null*† and the *power of the continuum*, and designated by $\aleph_0$ and c, respectively; they are the important transfinite cardinals. Only a very few mathematicians ever study any other infinities explicitly. However, in many branches of modern mathematics expounded at the college-graduate level (abstract algebra, Abelian group theory, homological algebra, as well as general topology), the trend is toward theorems and proofs that are valid for systems of large but unspecified cardinality.

A generation ago the logical soundness of such theorems was the subject of strong controversy. The possibility that some infinite sets represent a more abundant infinity than others had not been seriously imagined before Cantor's work in the 1870's. And yet, the definitions he adopted and the proofs he created are very simple, requiring almost no mathematical training for their comprehension.

---

† Aleph ($\aleph$) is the first letter of the Hebrew alphabet.

DEFINITION. Two sets (whether they are finite or infinite) are said to represent *the same cardinal number* if the elements of one set can be paired off against the elements of the other so that no elements are left over in either of the two sets.

For example, the even integers and the odd integers can be paired off in this way:

$$(1, \ 2), \qquad (3, \ 4), \qquad (5, \ 6), \qquad \cdots, \qquad (2n \ - \ 1, \ 2n), \qquad \cdots$$

and it follows that there are just as many of the one as of the other. Such a pairing-off is called a *one-to-one correspondence*.

The set of all integers can be put into one-to-one correspondence with the set of odd integers, for example, by the formula

$$N = 2n - 1,$$

where n ranges over the set of all integers, and the corresponding odd number is N. Thus the set of odd numbers (and also the set of even numbers) has the cardinal power $\aleph_0$ of the set of integers.

Many sets which are apparently more numerous than the integers are also of power $\aleph_0$, for example, the set of all rational numbers. It is easily seen that the rational numbers can be arranged without omission or repetition in a simple sequence. For instance, by first lumping together in groups all those rational numbers a/b which have $a + b = 2$ (i.e., $\frac{1}{1}$), then those which have $a + b = 3$ (i.e., $\frac{1}{2}, \frac{2}{1}$), etc., and then ordering the numbers in each group according to the size of their numerators, we have:

$$\frac{1}{1}, \qquad \frac{1}{2}, \frac{2}{1}, \qquad \frac{1}{3}, \frac{3}{1}, \qquad \frac{1}{4}, \frac{2}{3}, \frac{3}{2}, \frac{4}{1},$$

$$\frac{1}{5}, \frac{5}{1}, \qquad \frac{1}{6}, \frac{2}{5}, \frac{3}{4}, \frac{4}{3}, \frac{5}{2}, \frac{6}{1}, \qquad \text{etc.}$$

The only troublesome detail here is that one must avoid the repetition of the (equivalent) fractions in order to conform to the requirement of a one-to-one correspondence. For example, we have omitted $\frac{2}{2}$ from the third group because its equivalent, $\frac{1}{1}$, already appears in the first.

The fact that the set of rational numbers has the cardinal power $\aleph_0$ is a special case of the following:

THEOREM. *Let S be a set of sets, $S_1, \quad S_2, \quad S_3, \quad \cdots$. If S is finite or has the power $\aleph_0$, and if all sets $S_1, \quad S_2, \quad S_3, \quad \cdots$ have the power $\aleph_0$, then the set of all objects belonging to these sets also has the power $\aleph_0$.*

This assertion can also be written in the form of equations:

$$1 \cdot \aleph_0 = \aleph_0 , \qquad 2 \cdot \aleph_0 = \aleph_0 ,$$
$$3 \cdot \aleph_0 = \aleph_0 , \qquad \cdots , \qquad \aleph_0 \cdot \aleph_0 = \aleph_0^2 = \aleph_0 .$$

The proof is simple. By hypothesis, every set S_1, S_2, S_3, $\cdots$ can be arranged in a sequence. Let us denote by $x(i, j)$ the ith object in S_j. For instance, $x(2, 3)$ denotes the 2nd object in the 3rd sequence S_3. Now there is only a finite number of terms $x(i, j)$ with $i + j$ equal to some number a. Hence we can arrange the terms with $i + j = 2$ into a sequence, follow it by the sequence containing all terms with $i + j = 3$, follow this by a sequence containing all terms with $i + j = 4$, etc. Then we obtain a sequence containing each object occurring in any S_j *at least* once. Strike out the repetitions and you have a sequence containing each of our objects *exactly* once.

When a set of objects is put into one-to-one correspondence with the integers in this way, it is said to be *counted*. A set which can be counted is called *countable*. The theorem stated above asserts that a finite or countable infinity of countable infinities is countable. A set which cannot be put into one-to-one correspondence with the integers in this way is called *uncountable*.

We have shown how to count the rational numbers, and the next indicated step is to count the reals. But this is impossible. Cantor proved the following celebrated theorem:

THEOREM. *Given a one-to-one correspondence between the integers and some set of real numbers, it is always possible to construct a real number which has not been counted in this correspondence.*

COROLLARY. *The set of all real numbers is uncountable.*

The proof is simple; let us first concentrate on a convenient subset S of the real numbers: S denotes the set of all real numbers between 0 and 1 whose non-terminating decimal expansions use only two digits, say 1 and 2.

Let us suppose that we are given a one-to-one correspondence between the integers and the elements of S. This would mean that we can count the numbers in S, i.e., that we can make a list and refer to the first, second, $\cdots$ numbers in the list. We shall construct a number N belonging to S and we shall show that N is not in our list. This will prove that the subset S of real numbers is not countable and hence that the whole set of real numbers is not countable.

To construct N, select its first digit by looking at the first number in the list. If it has 1 as its first digit, let N have 2 as its first digit and vice versa. Similarly, select the second digit of N so that it is different from the second digit of the second number in the list. Proceed in this way, always making the kth digit of N different from the kth digit of the kth number in the list. Clearly, the number N so constructed is different from the first number in the list (because of its first digit), from the second number (because of its second digit), and, in general, from the kth number (because of its kth digit), and this is true for all k. So the real number N belongs to S (since it consists only of digits 1 and 2) but was not counted in our one-to-one correspondence. This concludes the proof.

This wonderful proof becomes clearer if one sets up some sequence of real numbers and applies the construction to obtain a number not in the sequence. Used this way, the argument in Cantor's proof is called *Cantor's diagonal procedure* and it is a standard way of constructing a number which differs from every one of a given infinite sequence of numbers.

Problem

3.14. We showed above how to count the rational numbers. Since these are also real numbers, they may serve as the set S in the preceding proof. Cantor's construction then yields a number which is not rational. Look into this argument, working out the first dozen or so digits of the resulting number N and using a variant of the scheme indicated above. (Make every rational number in the list on page 54 non-terminating, either by writing out all the zeros or by converting the terminating decimals to numbers ending in a string of nines, as you wish.)

3.10 Uses of the Uncountable

The distinction between countability and uncountability has turned out to be one of the cornerstones of analysis and topology. Here a few facts will be cited, without proof, to show some ways in which it can matter whether a given set is countable or uncountable.

Figure 3.16

First, one can have a countable infinity of segments on a line, all of the same size, no two of which touch each other; see Figure 3.16. One sees how to construct a countable infinity of segments on a given segment of length L, no two of which touch (Figure 3.17). Of

course, in this second case, only a finite number of them can be
larger than a preassigned unit fraction $1/n$. More precisely, there
must be fewer than nL segments of length at least $1/n$.

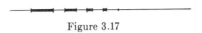

Figure 3.17

By contrast, in any uncountable infinity of segments on a line,
some must overlap. For every segment contains some rational points,
i.e., points whose distance from a fixed point 0 is a rational number.
Some rational points must belong to more than one segment, since
otherwise the set of segments could be counted, contrary to assumption. The same argument shows that in any uncountable set of segments on a line, uncountably many must overlap and uncountably
many must be larger than some unit fraction. More of this sort of
argument appears at the end of Chapter 6.

An amusing variation of this same phenomenon, which is actually
important in certain topological applications, is the following: Let us
suppose that the letters of the alphabet are represented by mathematical curves—that is, they are without any thickness. Also, suppose that the letters have no ornaments, i.e., no extra lines or feet.
Then one can scribble an infinity of each letter on any sheet of paper
of any given size. One can find room for an uncountable infinity of
certain letters, like L or C or I or M or N, but surprisingly it is
not possible to write an uncountable infinity of T's or A's or B's
or P's without two of them overlapping.

Problem

3.15. Can you classify now the letters of the alphabet according to the principle here indicated, and get to the topological essence of this fact?
(See Figures 3.18, 3.19 and 3.20.)

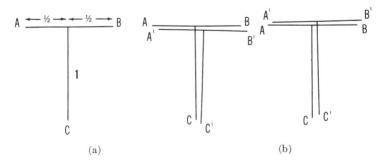

Figure 3.18

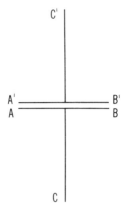

Figure 3.19

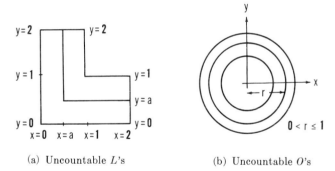

(a) Uncountable L's   (b) Uncountable O's

Figure 3.20

CHAPTER FOUR

Zig-zags: To the Limit if the Limit Exists

In this chapter we shall discuss the concept of *sequential limit*. This notion is much the same whether on the line, or in the real number system, or in the plane. We shall treat the case of the plane, because it contains a more stimulating variety of situations, and because one can draw illuminating pictures. Our principal tool for the exploration of the limit concept will be the *zig-zag*. This is an infinite sequence of line segments forming a simple polygonal path looking like the conventional representation of thunderbolts; see Figure 4.1. A picture of a thunderbolt suggests a target and this is a good approach to the idea of a limit. There is another point to the analogy, if we recall that anticipated limits sometimes do not exist and then realize that this is also true of supposed targets.

We shall find that when the target of a zig-zag is clearly marked, it will always be that point of the plane which is also the limit point of the sequence of vertices (corner-points) of the zig-zag.

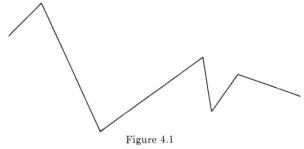

Figure 4.1

The zig-zags are interesting figures, especially in connection with the concept of *length*.

Figure 4.2

The zig-zag in Figure 4.2 clearly has no target; moreover, if continued indefinitely, it is infinitely long. The zig-zag in Figure 4.3 starts at P_1 ; for any finite n, the left part $P_1P_2 \cdots P_n$ is finite (the three dots in the middle just indicate that its exact shape does not matter here). The three dots on the right mean that the zig-zag has infinitely many vertices. The zig-zag is an infinite construction; for each $n = 1, \ 2, \ 3, \cdots$, there is a path $P_1P_2 \cdots P_n$ from the first vertex P_1 to the nth vertex P_n and the length of this path (let us call it L_n) is the sum of the lengths of the constituent segments. Thus

$$L_n = P_1P_2 + P_2P_3 + \cdots + P_{n-1}P_n .$$

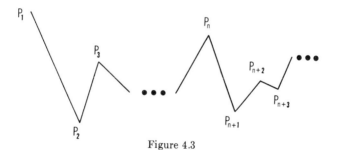

Figure 4.3

Now we make the following definition:

DEFINITION. The *length* L of the zig-zag $P_1P_2P_3 \cdots$ is the limit of the lengths L_n as n increases indefinitely, if this limit exists. Symbolically,

$$L = \text{length } (P_1P_2P_3 \cdots) = \lim_{n \to \infty} (\text{length } P_1P_2 \cdots P_n);$$

or, even more briefly, $\qquad L = \lim_{n \to \infty} L_n .$

Problem

4.1. What necessary qualifying phrase appears finally to have been omitted? How therefore must this last symbolism be read?

In general, as we see from the definition of this limit, in order to calculate it, we must sum an infinite series. We have had some practice in this already.

The first examples which follow are mainly concerned with *length*. The point T, origin of coordinates and undoubted target of the zigzag, will not figure importantly for the time being.

EXAMPLE 1. The first zig-zag consists of the segments A_0A_1, A_1A_2, A_2A_3, $\cdots$. The points A_n, $n = 1$, 2, 3, $\cdots$ are alternately above and below the x-axis; the abscissa of A_n is $1/2^n$, and the ordinate is the same except for sign. See Figure 4.4.

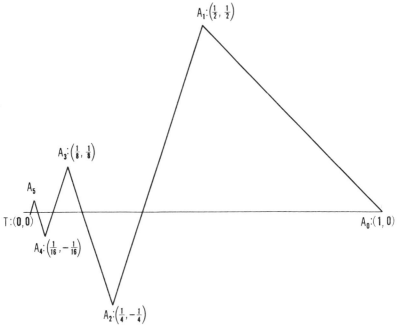

Figure 4.4

A_n is the point $\left(\dfrac{1}{2^n}, \dfrac{(-1)^{n+1}}{2^n} \right)$, $n = 1$, 2, 3, $\cdots$.

The zig-zag $A_0A_1A_2 \cdots$ is like a *half-line* or *ray*, and it is also like a segment with just one endpoint (corresponding to A_0), the other one having been deleted. Such segments occur frequently in mathematical work and are called *half-open*; the segment with both endpoints off is called *open* and the segment with both endpoints put back on is called *closed*. Notice that as far as the construction goes, the zig-zag does not contain the point T (although it obviously points to it).

It may surprise the reader that this infinite zig-zag is of finite length. It is easy to see that the length of any path $A_0 A_1 \cdots A_n$ along the zig-zag is *less* than the following sum which is itself less than 3:

$$1 + 2 \cdot \left(\frac{1}{2} + \frac{1}{4} + \frac{1}{8} + \cdots + \frac{1}{2^n} \right).$$

We get this sum if we replace each segment of the zig-zag by its horizontal *plus* vertical projections. Using the theorem of Pythagoras we can readily show that

$$A_0 A_1 = \frac{\sqrt{2}}{2}, \qquad A_1 A_2 = \frac{1}{4} \sqrt{10},$$

$$A_2 A_3 = \frac{1}{8} \sqrt{10}, \qquad A_3 A_4 = \frac{1}{16} \sqrt{10}, \qquad \cdots$$

and prove that the length L of the zig-zag is $\frac{1}{2} \sqrt{2} + \frac{1}{2} \sqrt{10}$. Thus $L \sim 2.3$ with an error as large as .2, i.e., an error not exceeding 10 per cent.

EXAMPLE 2. In Figure 4.5 the length is even easier to calculate. TB_1 is of unit length and we are given an infinite sequence of equilateral triangles, with bases of diminishing lengths: $\frac{1}{2}$, $\frac{1}{4}$, $\frac{1}{8}$, $\cdots$.

Since the bases add up to $TB_1 = 1$, one sees easily that the length of the zig-zag is 2.

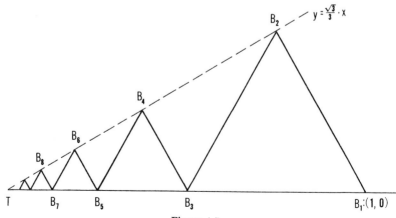

Figure 4.5

Problem

4.2. Notice that the odd-numbered vertices lie on the x-axis (where $y = 0$ in rectangular coordinates) and the even-numbered vertices lie on the line TB_2, whose equation is easily found to be

$$y = \left(\frac{\sqrt{3}}{3}\right) \cdot x.$$

Construct an analogous zig-zag (segments of equilateral triangles) whose even vertices lie on the curve $y = x^2$. How does the length of this zig-zag compare (judging from the pictures) with that of the preceding zig-zag?

EXAMPLE 3. The zig-zag of Figure 4.6 does not have finite length, and we shall say that it has infinite length in accordance with the following definition:

DEFINITION. If the lengths L_n increase without bound as n increases, then we say that L is *infinite*.

Since we are adding positive numbers, the numbers L_n must get bigger and bigger; if they had an upper bound (any number which is bigger than all of them), then, by the Bolzano-Weierstrass principle, they would have a limit (less than or equal to this bound). Therefore, if there is no limit $\lim L_n$, there is no bound, and that is all we mean to say when we call the length L infinite. In much the same way, one says that the length of a Euclidean line is infinite. It is a new form of words, a matter of convenience; *this* infinity is *not a number* of the real number system. Mathematicians sometimes augment the real number system, throwing into it a new object, ∞, to correspond to this infinity, but then they have to make special rules for operating with it. For example: $\infty + \infty = \infty$ but $\infty - \infty$ is *indeterminate*.

The zig-zag of Figure 4.6 is very much like the preceding one in appearance, but the coordinates of the vertices are different.

$$C_n \text{ is the point } \left(\frac{1}{n}, \frac{(-1)^n}{n}\right), \qquad n = 2, 3, 4, \cdots.$$

It is clear that the length of C_1C_2 exceeds $\frac{1}{2}$, that of $C_1C_2C_3$ exceeds 1, and that of $C_1C_2C_3C_4$ exceeds 1.5. This suggests that the length of $C_1C_2 \cdots C_n$ increases substantially with each step; but this is not true. The increase in later steps is very small. However, these lengths do add up and L_n increases without bound. Let us look into it more closely.

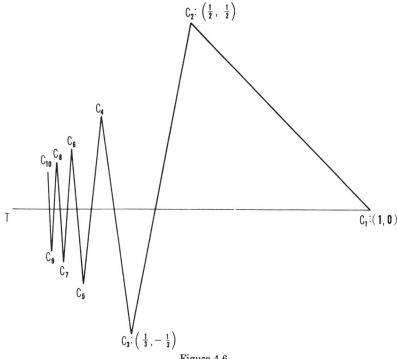

Figure 4.6

If we consider the four segments C_4C_5, C_5C_6, C_6C_7, C_7C_8, we see that the ordinate of each endpoint is, numerically, at least $\frac{1}{8}$, and therefore each segment has a length of at least $2 \cdot \frac{1}{8} = \frac{1}{4}$. Thus the four segments have a combined length of at least 1. In precisely the same way one sees that the next eight segments have a combined length of at least 8 times $\frac{1}{8}$, or 1, and the next sixteen segments after those have a total length of at least 1, and then the next 32, and so on.

In this way we can make a sequence of paths,

$$C_1C_2 \cdots C_8, \qquad C_1C_2 \cdots C_{16}, \qquad C_1C_2 \cdots C_{32}, \qquad \cdots$$

whose lengths increase like the numbers 3, 4, 5, $\cdots$ and since we can continue this indefinitely, we see that the lengths L_n increase without limit. The zig-zag has infinite length.

This result is of such great importance in mathematics that the reader will do well to think it through again, verifying that the length

$$C_NC_{N+1} + C_{N+1}C_{N+2} + \cdots + C_{2N-1}C_{2N}$$

exceeds 1, whenever N has the form 2^n. A convenient way to analyze this problem for oneself is to study the following example:

$$1 + \frac{1}{2} + \frac{1}{3} + \frac{1}{4} + \cdots .$$

This is called *the harmonic series*. It is a divergent series, that is, the partial sums

$$1 + \frac{1}{2} + \frac{1}{3} + \cdots + \frac{1}{n}$$

increase without limit as n increases.

A sequence of terms with indicated additions (or subtractions) is called a *series*. By definition, the sum of the first n terms of a series ($n = 1,\ 2,\ 3,\ \cdots$) is called *the nth partial sum*, written S_n and the sum S of the series is $\lim_{n \to \infty} S_n$ if this limit exists. If all the terms are positive and no limit exists, then one (sometimes) speaks of the sum as infinite. In general, when no limit S exists, the series is called *divergent*; when the limit exists, the series is called *convergent*. When the limit exists for the given series but fails to exist for the series which has the same numerical terms, all made positive, then the given series is called *conditionally convergent*. For example, the series

$$1 - \frac{1}{2} + \frac{1}{3} - \frac{1}{4} + \cdots$$

is conditionally convergent; that is, it is convergent as it stands but not convergent when all the terms are made positive.

We shall now use zig-zags to explore the notion of limit.

DEFINITION. The point P is called a *limit* (short for *sequential limit point*) of a sequence of points $P_1,\ P_2,\ P_3,\ \cdots$ provided that the sequence of distances $P_1P,\ P_2P,\ P_3P,\ \cdots$ approaches zero. (Cf. definition of "a sequence of positive numbers approaches zero", Section 3.5.)

As the pictures show, the point T in our examples is a limit of the zig-zag vertices. The distances from T to these vertices† are easily calculated. In the first example,

$$TA_n = \sqrt{\left(\frac{1}{2^n}\right)^2 + \left(\frac{1}{2^n}\right)^2} = \frac{\sqrt{2}}{2^n}.$$

In the second example, let the reader calculate TB_n (there being two cases to consider).

† Here we mean the length of the segment TA_n, *not* distance along the zig-zag.

Problem

4.3. Look at Example 2 in this roundabout way: First, show that the sequence B_1, B_3, B_5, $\cdots$ has T as limit. Next, use the fact that the distances $B_{2n}B_{2n-1}$ approach zero.

It follows easily from the triangle inequality† that a sequence cannot have more than one limit. For if the distances P_nT and also P_nT' approach zero, then the formula

$$P_nT + P_nT' \geq TT', \qquad n = 1, \ 2, \ 3, \cdots$$

shows that the distance TT' must equal zero. But in Euclidean geometry this is only possible when T and T' are the same point. Thus if we know that a limit exists we may speak of *the* limit of a sequence.

The next pictures are directed to the notion of limit point.

EXAMPLE 4. The zig-zag $OD_1D_2D_3 \cdots$ in Figure 4.7 uses segments which make an angle of $45°$ with the horizontal and therefore make right angles at each vertex. The length of D_mD_{m+1} is $1/2^m$, for $m = 1, \ 2, \ 3, \cdots$.

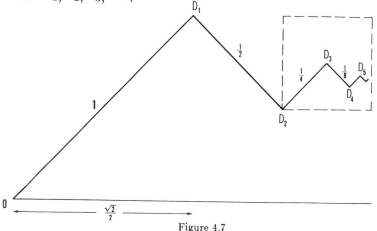

Figure 4.7

The infinite zig-zag starting at D_2, namely $D_2D_3D_4D_5\cdots$, lies entirely inside a square whose diagonal has length $\frac{1}{2}$ (sides are $\sqrt{2}/4$), it has D_2 as endpoint and no other point which we have yet

† This states that, in a triangle, the sum of the lengths of two sides is greater than the length of the third.

constructed is an endpoint of it. *However*, there is a point in the plane very intimately related to this zig-zag. Let us call it Z. Figure 4.7 shows Z where it does not belong; if the reader agrees that it does not belong there, then he will know how to find where it does belong.

Problem

4.4. Suppose that the point 0 is the origin of a coordinate system whose x-axis is horizontal as customary. Locate the point whose abscissa is $\sqrt{2}$ and whose ordinate is

$$\frac{\sqrt{2}}{2}\left(1 - \frac{1}{2} + \frac{1}{4} - \frac{1}{8} + \cdots\right) = \frac{\sqrt{2}}{2}\cdot\frac{2}{3} = \frac{\sqrt{2}}{3}.$$

Call this point Z and discuss the relation of Z to the vertices D_1, D_2, D_3, $\cdots$; explain what is meant by the statement "the abscissa of Z is a *limit of the sequence of abscissas* of the infinite succession of vertices". Prove that the ordinate of Z is a limit of the sequence of ordinates of the vertices (recall the formula for the sum of an infinite geometric progression).

EXAMPLE 4'. The zig-zag $OD'_1D'_2D'_3\cdots$ of Figure 4.8 is an interesting companion piece to the preceding. The diagram shows the same sequence of segments arranged to make a spiral.

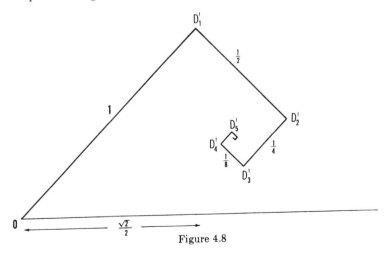

Figure 4.8

Problem

4.5. The reader may enjoy proving that there is a limit point T to the set of vertices D'_1, D'_2, $\cdots$ of Example 4' by finding its coordinates.

68   USES OF INFINITY

EXAMPLE 5. Let us look next at an infinite zig-zag $OE_1E_2E_3E_4 \cdots$ whose successive segments have lengths

$$1, \quad \frac{1}{2}, \quad \frac{1}{3}, \quad \frac{1}{4}, \quad \frac{1}{5}, \quad \cdots .$$

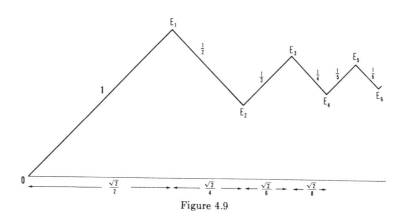

Figure 4.9

Its picture in Figure 4.9 makes all details clear. This example is similar to the zig-zag $OD_1D_2D_3 \cdots$ of Example 4, at least to the unaided eye. However, we already know something about adding the sequence of unit fractions: *They do not have a finite sum*. This will have striking consequences.

One might guess that this picture has a point Y associated with it, in the way in which the point Z of Example 4 is associated with its zig-zag. BUT such a guess is incorrect. This zig-zag moves to the right slowly but invincibly; *no point of the plane is a limit point of the sequence of points* E_1, E_2, E_3, E_4, $\cdots$. This is proved by the fact that the horizontal projection of the path $E_1E_2E_3 \cdots E_n$ has length

$$\frac{\sqrt{2}}{2} \left(1 + \frac{1}{2} + \frac{1}{3} + \frac{1}{4} + \frac{1}{5} + \cdots + \frac{1}{n} \right),$$

and this sum *increases without limit as* n *increases*. Thus, no matter what point Y one may choose, some E_n (for sufficiently large n) is to the *right* of this chosen Y; all later points of the zig-zag are even further to the right and cannot approach Y. The details should now be clear.

Problems

4.6. In this example, although the abscissas of the points E_n increase without limit as n increases, the *ordinates* do not increase in this way. Prove that the ordinate of E_n is

$$S_n = \frac{\sqrt{2}}{2} \left[1 - \frac{1}{2} + \frac{1}{3} - \frac{1}{4} + \frac{1}{8} - \cdots + \frac{(-1)^{n+1}}{n} \right],$$

and satisfy yourself that the sequence of ordinates has a limit by relating this problem to the picture in Figure 4.10.

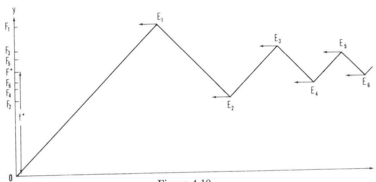

Figure 4.10

4.7. (a) Explain why the following assertion might be false, and formulate a correct statement: "The limit of the projections on a line (of an infinite sequence of points in the plane) is also the projection of the limit."

(b) Can you prove the following assertion? "The projection on a line of the limit (of a sequence of points) is the limit of the projections."

Two companion pieces to the previous zig-zag, the first showing convergence and the second divergence are pictured in Figure 4.11.

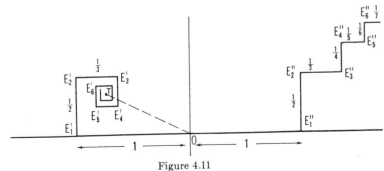

Figure 4.11

In the first one, the vertices have a limit point T, and the sequence of segments OE'_n ($n = 1$, 2, 3, $\cdots$) has a limiting position OT. In the second there is no limit point, but the sequence of segments OE''_n ($n = 1$, 2, 3, $\cdots$) does have a limiting direction! The reader is invited to prove that the limit of the slopes of OE''_n exists.

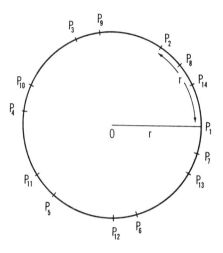

Figure 4.12

That π is irrational means that the radius of a circle and its perimeter are incommensurable lengths. This has a striking consequence, related to the present discussion. If we choose a point, call it P_1, on a circle and then rotate the circle through one radian we will get a new point, call it P_2 (the arc P_1P_2 has length equal to the radius), and if we rotate again through one radian, we will get another point P_3, and so on. Let us in this way construct a sequence of points P_n, $n = 1$, 2, 3, $\cdots$; see Figure 4.12. Now the points in this sequence can never repeat themselves, because of the incommensurability of the radius and perimeter. (This is not trivially obvious, and is worth some thought.) The sequence of radial segments OP_n cannot assume a limiting position, because we are constantly changing position by a whole radian. It follows from both these facts that the points P_1, P_2, P_3, $\cdots$ are distinct from each other, and that the sequence has no limit point.

The behavior of the point P_n as n increases is a very important problem in the higher calculus. It is easy to conjecture that P_n moves around so as to make its presence known in every part of the circumference more or less equally often, and this is the substance of a difficult and celebrated theorem due to Kronecker. The easier part of this theorem states that the set of points P_n, $n = 1$, 2, 3, $\cdots$ is *everywhere dense* on the circle. This means that if we pick some point P on the circle we can always find an *iterate* of P_1 (that is a point obtained from P_1 by a certain number of rotations through one radian) which is near to it. More precisely, if we choose some (large) integer, say K, we can find a point P_N (for some clever choice of N) so that the distance PP_N is smaller than $1/K$. Of course, this does not mean that P is a limit point of the entire sequence P_1, P_2, P_3, $\cdots$. Not at all; the very next point P_{N+1} will be at a substantial distance from P and we may have to go round the circle quite a few more turns, perhaps M turns, with some clever choice of a point P_{N+M}, before coming in as close to P as we earlier specified, namely $1/K$.

We shall close this album of illustrations of the limit concept with a remark on the Kronecker Theorem. If we choose an angle of rotation equal to k radians and consider the resulting sequence of points

$$P_1', \quad P_2', \quad P_3', \quad \cdots,$$

two cases are possible. If k is incommensurable with π we get an everywhere dense set, just as in the case above where $k = 1$. If k is a rational part of 2π, i.e., $k = 2m\pi/n$ with m and n integers, then the sequence is periodic and has the form

$$P_1', \quad P_2', \quad \cdots, \quad P_N', \quad P_1', \quad P_2', \quad \cdots.$$

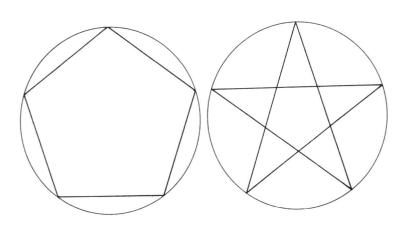

Figure 4.13

The number N of distinct points is at most n (remember that 2π radians is a full rotation), and N is always a factor of n (this is not easy to prove, but it does not take many lines; this result is one of the fundamental theorems in elementary group theory). This alternative gives rise to all the regular polygons and to a variety of beautiful pictures. Here, we illustrate the case $n = 5$; for $k = 2\pi/5$, we get the first, for $k = 4\pi/5$, the second of the polygons shown in Figure 4.13.

In all cases but one, a periodic set does not have a sequential limit point. The exceptional case is illustrated by a sequence Q, Q, Q, $\cdots$ and has Q as limit. A numerical instance is the sequence 2, 2, 2, $\cdots$, which has the limit 2. The idea behind this may be seen in the representation of 2 as a non-terminating decimal:

$$2.00000 \cdots ;$$

when we "round" this to the nth decimal, we get the nth approximation, which is still 2, of course, but we now view it as an approximation to 2 (which just happens to have hit the mark exactly). This is entirely in accordance with the definition of limit given above.

A final remark about the use of limits in the calculus. The notions of tangent to a curve, length of arc, area bounded by a curved figure, area of a piece of surface, curvature of a surface, volume bounded by a surface, center of gravity of a mass, and the physical notions of velocity, acceleration, work done, total gravitational attraction of one mass upon another—all of these and literally hundreds of others in every branch of the mathematical sciences—begin with definitions involving limits. The calculus works with all of this material.

In this chapter we have made one application of the notion of limit, namely to the calculation of the lengths of zig-zags. This is closely related to the mathematical problem of *summing infinite series*, one of the two most important uses of the limit notion in the calculus (the idea of a *derivative* associated with the slope of a tangent line to a curve being the other). The infinite series is a generalization of the notion of a non-terminating decimal such as

$$\frac{1}{3} = 0.3 + 0.03 + 0.003 + \cdots$$

and

$$\pi = 3 + 0.1 + 0.04 + 0.001 + \cdots .$$

(In the second example, the three dots merely mean that the expansion is non-terminating; no simple rule is implied for calculating the next digit.)

Problems

4.8. The following problem introduces the famous series

$$1 + \frac{1}{2^2} + \frac{1}{3^2} + \cdots + \frac{1}{n^2} + \cdots$$

which is convergent, whereas the harmonic series is not. It is proved in the calculus that the sum is $\pi^2/6$. All that the reader is asked to show is that the sequence of nth partial sums is bounded by 2. This will prove that certain points in the plane, about to be defined, do not "move off to infinity."

4.9. Figure 4.14 is similar to the "Whirl of Irrationals" (see Figure 3.10); but here, the legs of the nth triangle are related to the harmonic series $1, \frac{1}{2}, \frac{1}{3}, \cdots$ as follows: the segment P_nP_{n+1} $(n = 1, 2, 3, \cdots)$ is of length $1/n$ and it is perpendicular to the radial line OP_n. Now prove that, for all n, OP_n is less than $\sqrt{3}$. Next verify that the length of zig-zag $P_1P_2P_3 \cdots$ is infinite. Imagine the points P_n projected radially onto a convenient circle, say of radius 3, and denote the projection of P_k by Q_k. Can you prove that these points $Q_1, Q_2, Q_3, \cdots$ wind around indefinitely often? Can you conclude that there is no limit to the original sequence?

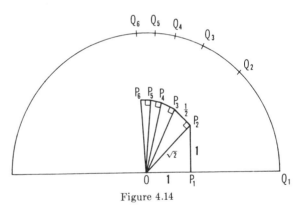

Figure 4.14

4.10. Determine whether or not the following series converge:

(a) $\frac{1}{2} + \frac{1}{4} + \frac{1}{6} + \frac{1}{8} + \cdots$

(b) $\frac{1}{3} + \frac{1}{7} + \frac{1}{11} + \frac{1}{15} + \cdots$

(c) $\frac{1}{1} + \frac{1}{\sqrt{2}} + \frac{1}{\sqrt{3}} + \frac{1}{\sqrt{4}} + \cdots$

(d) $\frac{1}{\sqrt{2}} + \frac{1}{\sqrt[3]{3}} + \frac{1}{\sqrt[4]{4}} + \cdots + \frac{1}{\sqrt[n]{n}} + \cdots$

4.11. Let P be a point and let P_n $(n = 1, 2, 3, \cdots)$ be a sequence of points in the plane. Suppose that for every line l, the projection of P on l is a limit of the projections of P_n $(n = 1, 2, 3, \cdots)$.

(a) Prove that P is then a limit of the sequence.

(b) Could this result be deduced from the fact that the given data are true for just one line?

(c) For two lines? What is the precise state of affairs?

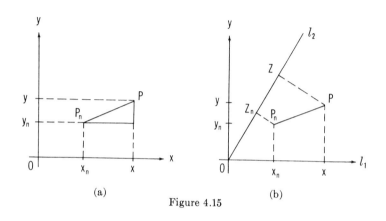

(a)   (b)

Figure 4.15

CHAPTER FIVE

The Self Perpetuating Golden Rectangle

In this chapter we shall study some of the properties of a rectangular shape discovered by the Greeks about 500 B.C., the time of the discovery of the irrationality of certain square roots. The construction of this rectangle involves $\sqrt{5}$, a number which is the key to the ruler-and-compass construction of the regular pentagon. This pleasing polygon is a face of the regular dodecahedron (twelve-sided polyhedron) and is built into the regular twenty-sided icosahedron, the most fabulous of the five regular solids. The golden rectangle therefore was highly prized by the Greeks for its own beauty and for the beauty of its family connections.

Looked at in the right way it gives rise almost instantly to the following sequence:

$$1, \quad 1, \quad 2, \quad 3, \quad 5, \quad 8, \quad 13, \quad 21, \quad 34, \quad 55, \quad 89, \quad 144, \quad \cdots .$$

This succession of numbers was invented by Leonardo of Pisa early in the 1200's in a problem of arithmetic about the breeding of rabbits (1 and 1 in his sequence represent the ancestor-rabbits) and it has come to be called the Fibonacci† series. Ever since, the connection between the rectangle and this series has been a source of mathematical discoveries, some of which will be shown in this chapter. By the eighteenth century it had been found that the rectangle was naturally associated with a logarithmic spiral, a shape connected with the growth of many natural objects like some molluscs and snails.

† See footnote to p. 20.

The enunciation by Fechner, the great psychologist, of his famous law: "The intensity of reaction to a stimulus varies as the logarithm of the intensity of the stimulus", brought to the rectangle and its built-in spiral a host of admirers who were not necessarily mathematicians. The importance of logarithms in psychological phenomena is attested by a variety of facts; the use of the decibel, for example, in measuring intensity. A biological phenomenon called "phyllotaxis" also exhibits the Fibonacci numbers.†

5.1 The Golden Rectangle

Figure 5.1 shows a rectangle shaped like a librarian's 3 x 5 card. If one folds such a card in the manner shown in Figure 5.2, there is left a rectangle whose sides are in the ratio 2 to 3, which is about 0.67 to 1.

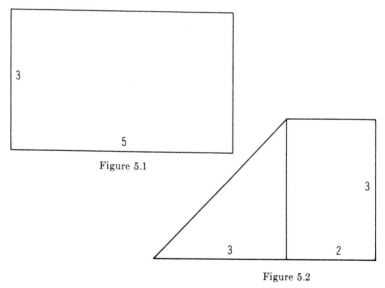

Figure 5.1

Figure 5.2

The original card had the ratio of 3 to 5 which is 0.6 to 1. Thus if one cuts a square away from such a card one gets a shape approximately like that of the original. Now the shape of the *golden rectangle* may be defined in this way: *If one cuts a square away from it then the rectangle that remains has exactly the same shape as the original rectangle.* By "same shape" we mean that the ratio of shorter side to longer side is the same, i.e., that *the two rectangles are similar.*

† For an account of these phenomena, see the beautiful discussion in H. S. M. Coxeter's *Introduction to Geometry*, Chapter 11. See also the article "Mathematical Games," by Martin Gardner, in *Scientific American*, August, 1959.

Clearly infinity is now let loose; for, if one cuts a square away from the newer rectangle, there results another similarly shaped rectangle, and if one cuts a square from this one there results another, and so on, indefinitely.

Let us denote the lengths of the sides of the original rectangle by $a + b$ and a. Then the lengths of sides of successive rectangles give the following sequence of pairs:

$$\begin{cases} a + b \\ a \end{cases}, \quad \begin{cases} a \\ b \end{cases}, \quad \begin{cases} b \\ a - b \end{cases}, \quad \begin{cases} a - b \\ 2b - a \end{cases},$$

$$\begin{cases} 2b - a \\ 2a - 3b \end{cases}, \quad \begin{cases} 2a - 3b \\ 5b - 3a \end{cases}, \quad \begin{matrix} 5b - 3a \\ 5a - 8b \end{matrix}, \quad \cdots,$$

where we have written the longer side of each pair first. The pattern which emerges is this: The longer side of one rectangle is the sum of the two sides of the next following rectangle. The coefficients of a and b in these expressions are related to the Fibonacci numbers listed above. The law of formation of the Fibonacci numbers is this: Each number is the sum of the two *preceding* numbers (exception being made for the two ancestors: 1 and 1). The laws of formation of the above lengths and of the Fibonacci numbers are so similar that one cannot doubt a strong connection; we shall come back to this later.

5.2. The Golden Mean is Irrational

The process of forming successive rectangles cannot end and this tells us that a and b are not commensurable (see Chapter 3) and therefore that $m = b/a$ is irrational. The historian of mathematics, Cajori, attributes the first proof of the irrationality of m, the *golden mean* (i.e. the ratio of the shorter to the longer side of a golden rectangle), to Campanus in 1260 "by means of a method which combined a reductio ad absurdum and the principle of infinite descent". Let us see how such a proof goes.

We suppose, for the sake of argument, that m is rational, that is, m is a ratio of some pair of integers. Then we may as well suppose that a and b are these integers and use Figure 5.3. Next, we see that

$$a - b, \quad 2b - a, \quad 2a - 3b, \quad 3a - 5b, \quad 5a - 8b, \quad \cdots$$

are all integers and the construction shows that they are all positive! Our assumption that m is rational has led us to an *infinite descending sequence of positive integers*. But this contradicts truth because there is no such sequence. Thus our assumption that m is rational is untenable and this concludes the proof that m is irrational.

The reciprocal of m, namely a/b, is often denoted by the Greek letter τ (tau), short for $\tau o\mu\eta$, "the section". Thus $\tau = 1/m$.

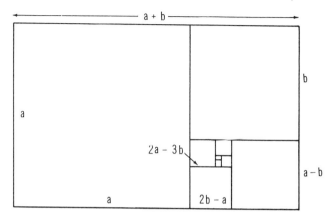

Figure 5.3

Problem

5.1. Use Figure 5.4 to prove the irrationality of $\sqrt{2}$ and Figure 5.5 similarly for $\sqrt{5}$. Can you adapt this to other cases of $\sqrt{n}$?

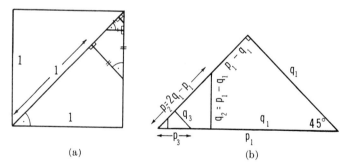

Figure 5.4

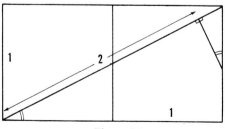

Figure 5.5

5.3. Estimating the Golden Mean

We have been assuming that a golden rectangle exists. This is easy to prove, but first let us estimate the golden mean by "trial and error". We wish to determine the ratio

$$m = \frac{b}{a} \quad \text{so that} \quad \frac{b}{a} = \frac{a}{a+b};$$

for, if a and b satisfy this relation, then the rectangle in Figure 5.3 is indeed a golden rectangle. If we choose a and b so that $a+b=1$, our calculations become simpler, for then

$$m = \frac{b}{a} = \frac{a}{a+b} = a;$$

that is, a and b must be so chosen that $m = a$ and $a+b = 1$. Accordingly, we make a table of trial sides a and b and record the ratio $b/a = m$. The first table indicates that m is between .6 and .8 near to .6; the second gives a better estimate of m, between .61 and .62. Therefore .615 cannot differ from the golden mean by more than 0.005 which is a possible error not exceeding 5 in 600 and is less than a 1 per cent error. This would be pretty good for manufacturing rectangular cards by machine methods.

a	.2	.4	.6	.8
b	.8	.6	.4	.2
m	4	1.5	0.67	0.25

a	.6	.61	.62
b	.4	.39	.38
m	0.67	0.65	0.61

Problem

5.2. m is nearer to .62 than it is to .61; use "proportional parts", or some simple graphical scheme for getting a better estimate for m. The value of m to nine decimals is 0.618 033 989, and a good graph should give 0.618.

5.4. Ways of Finding the Golden Mean

a) The shape appears to have been discovered in the following problem: "Divide a given line segment into two parts such that the shorter is to the longer as the longer is to the whole segment"; see Figure 5.6(a).
The conditions in the problem, expressed in symbols, are

$$\frac{b}{a} = \frac{a}{a+b}.$$

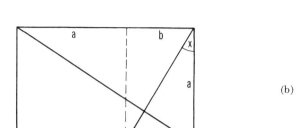

Figure 5.6

Dividing the numerator and denominator on the right by a, we obtain

$$\frac{b}{a} = \frac{1}{1 + \dfrac{b}{a}};$$

substituting m for b/a and denoting $1/m$ by τ, we get

$$m = \frac{1}{1 + m} \quad \text{or} \quad m(m + 1) = 1 \quad \text{or} \quad m + 1 = \tau.$$

This says that the reciprocal of m is $m + 1$. Also it leads to a quadratic equation and a precise expression for m (see below). But before deriving a formula for m, let us see the relation between this definition of m and the preceding one. Let us imagine that we have found the proper lengths a and b and let us express the equality of ratios by means of a pair of similar triangles. This may be done directly on the given segment as follows:

Figure 5.6(b) shows the similar right-angled triangles; two equal acute angles are marked by an x. Now the dotted lines complete the figure to a pair of similar rectangles and a residual square. Clearly we are back to the previous definition of the golden shape.

We now continue with the calculation of m. We saw that the definition of the golden mean implies that m satisfies the equation

$$m^2 + m = 1.$$

By completing the square:

$$m^2 + m + \frac{1}{4} = \frac{5}{4},$$

we find that

$$m + \frac{1}{2} = \frac{\sqrt{5}}{2} \quad \text{or} \quad m + \frac{1}{2} = -\frac{\sqrt{5}}{2} \, ;$$

since a and b are lengths, m must be positive, so

$$m = \frac{\sqrt{5} - 1}{2} \, .$$

This is the formula for m.

It is easy to check that m satisfies the original conditions because

$$m + 1 = \frac{\sqrt{5} + 1}{2} = \tau$$

and

$$m(m + 1) = \frac{\sqrt{5} - 1}{2} \cdot \frac{\sqrt{5} + 1}{2} = \frac{5 - 1}{4} = 1.$$

By looking up $\sqrt{5}$ in a table or by calculating it in one way or another, we can get arbitrarily close approximations to m; to nine decimals,

$$m \approx 0.618\ 033\ 989.$$

A ruler and compass construction is indicated in Figure 5.7, as the reader can verify. The length of the segment AE is 1, DE and FG are parallel, and the segments EG and GC have the desired length m. This is one way in which the Greek geometers solved the equation. The Babylonian mathematicians had worked out the theory of such quadratic equations for innumerable special cases (avoiding imaginaries and negatives) by 1800 B.C.

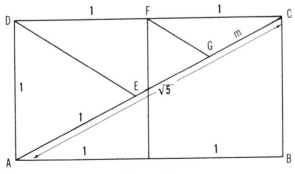

Figure 5.7

b) Another way, verging on the fantastic, of solving the equation

(5.1) $$m = \frac{1}{1+m}$$

was discovered by Girard, a mathematician of the 17th century. It consists of substituting, in the denominator of the right member of equation (5.1), the value of m, so that

(5.2) $$m = \frac{1}{1 + \dfrac{1}{1+m}},$$

and of repeating this substitution for m in each subsequent equation. Ordinarily, the object is to remove m from the right hand side of the equation; but this process pushes m out, way out. After a few steps, always by simple substitution, we see that

$$m = \cfrac{1}{1 + \cfrac{1}{1 + \cfrac{1}{1 + \cfrac{1}{1 + \cdots}}}}.$$

This formula is a very exciting statement of something, but of what precisely is hard to say.† One attempt to give it meaning is to *define* it as the limit of a sequence of compound fractions contained in the form above, namely:

$$\frac{1}{1}, \quad \cfrac{1}{1 + \cfrac{1}{1}}, \quad \cfrac{1}{1 + \cfrac{1}{1 + \cfrac{1}{1}}}, \quad \cfrac{1}{1 + \cfrac{1}{1 + \cfrac{1}{1 + \cfrac{1}{1}}}}, \quad \cdots.$$

The first five of these numbers are 1, $\frac{1}{2}$, $\frac{2}{3}$, $\frac{3}{5}$, $\frac{5}{8}$, and a moment's thought will guide the reader to the next one. Thereafter it is easy to guess, and even to show, that the integers appearing in the denominator of each (and in the numerator of the next) fraction are the Fibonacci numbers! Thus m is now given to us as a limit.

Problem

5.3. Show that if one of the fractions above is equal to p/q, then the next one must equal $q/(p+q)$.

† In order to find out what, precisely, is defined by such an expression, read *Continued Fractions* by C. D. Olds, to appear in this series.

The fact that these fractions actually converge to the number m, as one hopes they do, was proved by Simson (1734), 100 years after Girard's discovery. He showed that the successive fractions alternately overshot their target and undershot it; the even-numbered fractions constitute a decreasing sequence and the odd-numbered constitute an increasing sequence, both with limit m.

In the next section we shall see how this can be proved.

Problem

5.4. (a) Show that the following method of solving $m^2 = 1 - m$ is fantastic:

$$m = \sqrt{1 - m} = \sqrt{1 - \sqrt{1 - m}}$$
$$= \sqrt{1 - \sqrt{1 - \sqrt{1 - m}}} = \sqrt{1 - \sqrt{1 - \sqrt{1 - \cdots}}},$$

by showing that the sequence

$$\sqrt{1}, \qquad \sqrt{1 - \sqrt{1}}, \qquad \sqrt{1 - \sqrt{1 - \sqrt{1}}}, \qquad \cdots$$

of finite parts does not have m as limit.

(b) Show that τ, the reciprocal of m, satisfies the equation $\tau^2 = 1 + \tau$ and that the method of solution

$$\tau = \sqrt{1 + \tau} = \sqrt{1 + \sqrt{1 + \tau}}$$
$$= \sqrt{1 + \sqrt{1 + \sqrt{1 + \tau}}} = \cdots$$

is plausible because the corresponding sequence of finite parts has a limit.

5.5. Another Sequence Leading to m

In our present plan for approximating m, it will simplify matters to take a as unit of length, i.e., $a = 1$; then

$$\frac{b}{a} = \frac{b}{1} = m,$$

and the longer (also the shorter) sides of successive golden rectangles have lengths

$$1 + m, \quad 1, \quad m, \quad 1 - m, \quad 2m - 1, \quad 2 - 3m,$$
$$5m - 3, \quad 5 - 8m, \quad 13m - 8, \quad 13 - 21m, \quad \cdots .$$

It is geometrically clear that these numbers approach zero. But if

$5 - 8m$, for example, is small, i.e., if

$$5 - 8m \approx 0,$$

then m is approximately $\frac{5}{8}$, i.e., $m \approx \frac{5}{8} \approx .618$. This reasoning is quite general; each term set "approximately equal" to zero gives an approximation for m. Starting with the fifth, we get, in this way

$$\frac{1}{2}, \quad \frac{2}{3}, \quad \frac{3}{5}, \quad \frac{5}{8}, \quad \frac{8}{13}, \quad \frac{13}{21}, \quad \frac{21}{34}, \quad \frac{34}{55}, \quad \frac{55}{89}, \quad \cdots$$

Problem

5.5. Find the decimal equivalent of some of these. Try the ratios 377/610 and 610/987; how far out are these in the sequence?

The rule of formation of these fractions is clear: the denominator of one fraction is the numerator of the next and the *sum* of numerator and denominator gives the new denominator; that is,

$$\frac{p}{q} \quad \text{is followed by} \quad \frac{q}{p+q}.$$

It is immediately clear that the sequence of denominators is precisely the Fibonacci series, and of course this is also the sequence of the numerators; there is a small matter of the lost ancestors, 1 and 1, but it will be remembered we started without them.

Problem

5.6. Is each fraction above *numerically* closer to m than the preceding one? Give evidence in support of your answer.

We shall show that the successive approximations to m oscillate around it. Take the pair $\frac{34}{55}$ and $\frac{55}{89}$ as an example. These satisfy the simple relation:

$$(5.3) \qquad \frac{34}{55} \cdot \frac{55}{89} + \frac{55}{89} = 1.$$

It is not necessary to multiply out and find large common denominators in order to verify this; cancellation of the first pair of 55's shows the trick. Notice next the resemblance of this equation to the one satisfied by m if we write it in the form

$$(5.4) \qquad m \cdot m + m = 1.$$

Now let us suppose that $\frac{34}{55}$ is bigger than $\frac{55}{89}$ (as it is); if we replace $\frac{34}{55}$ by $\frac{55}{89}$ in (5.3) that equation will become false because the left side will become smaller. Thus $\left(\frac{55}{89}\right)^2 + \frac{55}{89}$ is smaller than 1. This shows that $\frac{55}{89}$ is smaller than m. Similarly supposing again that $\frac{55}{89}$ is smaller than $\frac{34}{55}$ and replacing $\frac{55}{89}$ in (5.3) by the larger number we find that $\frac{34}{55}$ is larger than m. It happens that $\frac{34}{55}$ *is* bigger than $\frac{55}{89}$, but even if it were not, what our argument shows is this: The smaller of these numbers is smaller than m and the larger of the two is larger than m. They cannot be equal in view of equation (5.4), as we shall now show.

The argument is quite general: A pair of successive approximations has the form p/q, $q/(p + q)$ (where p and q are integers) and satisfies the equation

$$(5.5) \qquad \frac{p}{q} \cdot \frac{q}{p + q} + \frac{q}{p + q} = 1.$$

If the two fractions could be equal, then comparison with equation (5.4) shows that each would be equal to m; but m is irrational (see Section 5.2). Therefore these fractions are not equal. To see that m lies between them, observe that if we replace the larger one by the other in equation (5.5), the left side will become smaller than 1; this shows [by comparison with (5.4)] that the smaller number is smaller than m. Similarly, the larger is larger than m. Q.E.D.

The reader will not have any trouble satisfying himself that the observations of Simson have now been verified. Before going over to the logarithmic spiral associated with the golden rectangle, we mention two interesting problems: one has to do with the powers of m; the other is a wonderful problem due to Lagrange and concerns the residues of the Fibonacci numbers modulo any integer. The second one is difficult. See the next chapter for its solution.

Problems

5.7. Extend the following remarks: $m^2 = 1 - m$; $m^3 = m - m^2 = 2m - 1$; $m^4 = 2m^2 - m = 2 - 3m$; $\cdots$. Can you write a general formula? Do the same for $\tau = 1 + m$; $\tau^2 = \tau + 1$; $\tau^3 = \tau^2 + \tau = 2\tau + 1$; $\cdots$.

5.8. (a) The Fibonacci numbers, modulo 2, are 1, 1, 0, 1, 1, 0, $\cdots$; modulo 3, they are 1, 1, 2, 0, 2, 2, 1, 0, 1, 1, 2, $\cdots$; both sequences are periodic. Show that the sequence of Fibonacci numbers mod 4 is also periodic. Similarly, exhibit its periodicity mod 5 and mod 6. (The series modulo 10 has period 60.)

(b) Do you see how to get the residues, mod 3, 4, 5, 6 without actually calculating the large Fibonacci numbers? Using this knowledge, and the rule that each number is the sum of the preceding two, can you prove Lagrange's remark: *The residues of the Fibonacci numbers modulo any integer whatever are periodic?* Hint: If n is the modulus the period does not exceed $n^2 + 1$.

5.6. A Spiral Zig-Zag

We shall next construct a spiralling zig-zag whose vertices lie on a very beautiful spiral. The zig-zag will lead us to the origin of this spiral, and also disclose the essential scheme underlying the construction of the spiral. At the same time we shall be constructing a whirl of contracting squares and also a whirl of rectangles and doing a bit of whirling ourselves. Starting at the lower left corner A of a golden rectangle $ABDF$, see Figure 5.8, we draw a 45° line to the upper edge at C, and drop the perpendicular CH cutting off a square. Next, using the new golden rectangle $CDFH$, with CH as a base, we repeat this construction; that is, we draw a line CE at 45° to the new base CH and drop a perpendicular EJ cutting off a square and giving us rectangle $EFHJ$ with base EJ. Then we draw a line EG so that EG makes a 45° angle with EJ, drop the perpendicular GK, and from the point G of the new rectangle $GHJK$ we continue the construction. In other words, every time we draw a line from the lower left vertex of a golden rectangle at 45° to the base, this line intersects the upper side of the rectangle in a point from which the next, similar, construction begins. Each new vertex belongs to a rectangle which is a 90° turn from the previous one and whose sides are m times those of the previous one. Each rectangle is whirled around by 90° and shrunk by a factor m.

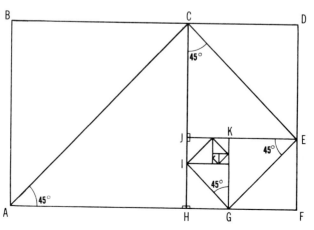

Figure 5.8

As far as the succession of squares and rectangles is concerned we have the same situation as before, and the same lengths of sides: $1 + m$, 1, m, $1 - m$, $2m - 1$, $\cdots$, and so on. In other words, we get the sequence m^{-1}, 1, m, m^2, m^3, $\cdots$, as the reader has already proved or can prove now merely by observing the figure (or from the governing equation $m^2 = 1 - m$), and the fact that each length comes from the preceding by scaling down in the ratio of 1 to m.

Problems

5.9. The whirl of rectangles may be expressed thus: $ABDF$, $CDFH$, $EFHJ$, $GHJK$, $\cdots$. Can you attach any significance to the way the vertices are ordered in successive rectangles?

5.10. If the side AB has length 1, AC has length $\sqrt{2}$. Show that the length of the zig-zag $ACEGI \cdots$ is $\sqrt{2}/m^2$ by using the formula for an infinite geometric progression. Have we proved that this formula is valid even if the common ratio is irrational?

5.7. Use of Similarity Transformations

It is clear from Figure 5.8 that the zig-zag closes down on some target-point T. We shall prove that T is the intersection of the diagonal BF of $ABDF$ and the diagonal DH of $CDFH$ (the second diagonal is the successor of the first under the whirling motion described above). The proof is technically very simple. In the preceding description we imagined ourselves turning through 90° with each new line of the zig-zag. Here we shall let the rectangles turn. Our goal is to place rectangle $ABDF$ on $CDFH$, A going into C, B into D, D into F, and F into H, by a combination of two motions, a rotation through 90° and a shrinking in the ratio 1 to m. We can describe the desired motion very simply, with T as the center of both motions. First we rotate the whole plane through an angle of 90° clockwise around T, so that the rectangle $ABDF$ goes into the new position $A'B'D'F'$; see Figure 5.9. Next we shrink all points of the plane towards T along radial lines in the ratio 1 to m, so that $A'B'D'F'$ goes into $CDFH$. This pair of motions, called transformations of the plane,† actually maps the points A, B, D, F upon the respective points C, D, F, H. The repetition of this composite transformation, indefinitely often, generates the whole zig-zag.

† In order to find out more about the subject of transformations, see *Geometric Transformations* by I. M. Yaglom (translated by Allen Shields) to appear in this series.

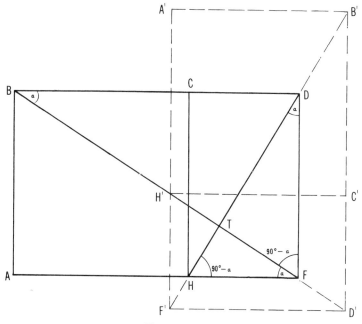

Figure 5.9

The first step in our proof consists in showing that the diagonal BF of the first rectangle meets the diagonal DH of the second rectangle at right angles at the point T. Since $ABDF$ and $CDFH$ are similar rectangles, all angles marked α in Figure 5.9 are equal. The similar right triangles BFD and DHF therefore have

$$\angle BFD = 90° - \alpha = \angle DHF,$$

so that

$$\angle HTF = 180° - [\alpha + (90° - \alpha)] = 90°,$$

and BF and DH indeed meet at right angles in T. Thus, if $ABDF$ is rotated by $90°$ around the point T, BF will go into a new position $B'F'$ perpendicular to the original BF (see Figure 5.9), and $B'F'$ will contain the line DH. Similarly, DH will go into $D'H'$ and $D'H'$ will lie along BF.

Next, observe that

$$\frac{B'D'}{DF} = \frac{1 + m}{1} = \frac{1}{m} = \frac{B'T}{DT} = \frac{D'T}{FT}.$$

Moreover, since

$$\frac{A'B'}{CD} = \frac{AB}{CD} = \frac{B'T}{DT} = \frac{1}{m},$$

and since CD and $A'B'$ are parallel, it is clear that C lies on $A'T$ and that

$$\frac{A'T}{CT} = \frac{1}{m}.$$

Similarly, H lies on $F'T$ and

$$\frac{F'T}{HT} = \frac{1}{m}.$$

Hence, the shrinking by a factor m of all distances from the point T brings the rectangle $A'B'D'F'$ into the rectangle $CDFH$ as we set out to show.

Our arguments depended only on the golden ratio m of the shorter to the longer side of the given rectangle, and not on its actual dimensions. Since this golden ratio is preserved in all the subsequent rectangles of our construction, the above proof is valid for all. This shows that the point T will serve as the center of rotation and contraction throughout, that it is interior to all the shrinking rectangles we construct by our scheme, and that it is therefore the target point of the zig-zag.

5.8. A Logarithmic Spiral

We can find a simple formula for the vertices A, C, E, G, I, $\cdots$, of the zig-zag in Figure 5.8 if we make the following choice of axes and units. Pick T as origin of coordinates, let r represent distance from T, and choose the line AT as initial direction (see Figure 5.10). Take the quarter-turn (one-fourth of a rotation $= \pi/2$ radians $= 90°$) as unit of angle, measured from AT, clockwise being taken positive, and let t denote the number of quarter-turns. The vertices A, C, E, G, $\cdots$ are obtained by rotating AT about T through 0, 1, 2, 3, $\cdots$ quarter-turns, respectively, and by shrinking the distance AT by m^0, m^1, m^2, m^3, $\cdots$, respectively. Therefore, the distance r of any such vertex from T of our spiral zig-zag can be written

(5.6) $$r = AT \cdot m^t.$$

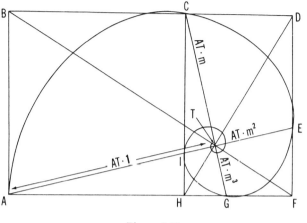

Figure 5.10

This gives the following table:

vertex	A	C	E	G	I	$\cdots$
t	0	1	2	3	4	$\cdots$
r	AT	$AT \cdot m$	$AT \cdot m^2$	$AT \cdot m^3$	$AT \cdot m^4$	$\cdots$
$\dfrac{r}{AT} = R$	1	m	m^2	m^3	m^4	$\cdots$

We get an even simpler formula if we take AT as unit of distance. We can do this by setting $R = r/AT$. Then, in terms of R, Equation (5.6) becomes

$$R = m^t.$$

If, for t, we substitute successively the terms of the arithmetic progression 0, 1, 2, 3, 4, 5, 6, $\cdots$, then we obtain, for R, the geometric progression 1, m, m^2, m^3, $\cdots$; the position of all these vertices is now determined because t is the number of quarter-turns and R is the factor by which AT is multiplied.

It can be shown that, if we now let t take on all positive real values (instead of merely integers) we get the smooth spiral pictured in Figure 5.10.

Problems

5.11. Calculate some in-between points on the spiral by using the values $\frac{1}{2}$, $\frac{3}{2}$, $\frac{5}{2}$, $\cdots$ for t. Find some additional points for other rational values of t.

5.12. What do we get if t also takes on negative values? How big does R become?

Once the curve is drawn it finds for us all powers of m; for example

$$m^{1/2} = \sqrt{m} \quad \text{corresponds to} \quad t = \frac{1}{2},$$

$$\text{that is,} \quad \frac{1}{2} \text{ quarter-turn} = \frac{1}{2} \ 90° = 45°;$$

$$m^{1/3} = \sqrt[3]{m} \quad \text{corresponds to} \quad t = \frac{1}{3},$$

$$\text{that is,} \quad \frac{1}{3} \text{ quarter-turn} = \frac{1}{3} \ 90° = 30°;$$

$$m^{5/6} = \sqrt[6]{m^5} \quad \text{corresponds to} \quad t = \frac{5}{6},$$

$$\text{that is,} \quad \frac{5}{6} \text{ quarter-turn} = \frac{5}{6} \ 90° = 75°.$$

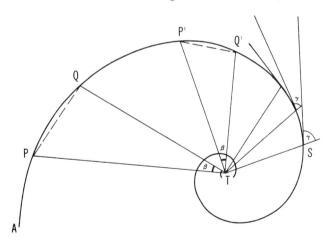

Figure 5.11

The curve

$$R = m^t$$

is called an *equiangular spiral* and also a *logarithmic spiral* (in this case the reference is to logarithms to the unusual base m). The equation can be put into other standard forms:

$$R = 10^{ct}, \qquad R = 2^{kt}, \qquad R = e^{ht},$$

by proper choices of the constants c, k, h. The advantage of this is that 10, 2, and e (≈ 2.718) are the more conventional bases for logarithms.

The logarithmic spiral has two important properties:

i) If points P, Q and P', Q' subtend equal angles (β) at T then the triangles PTQ and $P'TQ'$ are similar; see Figure 5.11.

ii) At each point (S) of the logarithmic spiral, there is a tangent line and it makes a constant angle (γ) with the radius vector TS.

Property i) is easy to prove and is left as an exercise. Property ii) is more difficult; one needs calculus to define the tangent to the curve, must show that this curve has a tangent everywhere (this is easy in this case provided one first shows that the curve has a tangent at some point), and then one can prove property ii).

Problems

5.13. Explain how the curve in Figure 5.11 can be used (equivalently to a table of logarithms) to multiply numbers if these are represented by lengths on a ruler which pivots at T *and* if we can measure and add appropriate angles.

5.14. What property of the curve corresponds to the use of the *characteristic* and *mantissa* of logarithms?

5.15. By plotting the curve and measuring with a protractor, check that the tangent to the curve at each point (S) makes an angle with the radius vector TS of about $73°$ [actually arc tan $(\pi/2 \log \tau)$].

5.9. The Pentagon

Perhaps because mathematicians work as hard as they do, the subject of mathematics rewards them with frequent bonuses. The logarithmic spiral generated by the shower of golden rectangles, and the appearance of the Fibonacci numbers in a sequence of approximations to the golden mean

$$m = \frac{\sqrt{5} - 1}{2}$$

are pleasant and unexpected encounters.

Another bonus of this sort is the fact that m is also the ratio of the side of a regular pentagon to its diagonal. In Figure 5.12(a) let the equal diagonals P_1P_3, P_2P_4, P_3P_5, P_4P_1, P_5P_2 be of unit length; then each side is m. A proof follows.

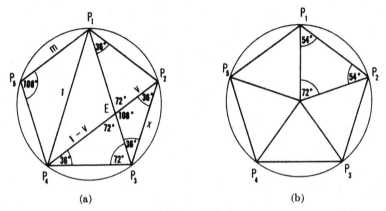

(a)   (b)

Figure 5.12

We define the regular pentagon by the condition that the arcs P_1P_2, P_2P_3, $\cdots$, P_5P_1 of the circumscribed circle are equal. This implies that the chords P_1P_2, P_2P_3, $\cdots$, P_5P_1 are equal. Each arc takes up one fifth of the circumference and subtends a central angle of $2\pi/5$ radians or $72°$; see Figure 5.12(b). There are five congruent central isosceles triangles whose angles are $54°$, $54°$, and $72°$, so each vertex angle of the pentagon is $108°$. Since triangles $P_1P_2P_3$, $\cdots$ [see Figure 5.12(a)] are isosceles, their base angles are

$$\tfrac{1}{2}(180° - 108°) = 36°; \qquad \angle P_2EP_3 = 180° - 2\cdot36° = 108°,$$

and

$$\angle P_1EP_2 = 72°.$$

The substantial part of the proof begins now. We find first that the triangles EP_2P_3 and $P_3P_4P_2$ in Figure 5.12(a) have angles $36°$, $36°$, and $108°$; they are isosceles and similar. Denoting EP_2 by v, P_2P_3 by x, and remembering that $P_2P_4 = 1$, we write the proportions:

$$\frac{v}{x} = \frac{x}{1}; \qquad \text{thus} \qquad v = x^2.$$

Next we find easily that the angles of triangle EP_4P_3 are $72°$, $72°$, $36°$; so that this triangle is isosceles and

$$EP_4 = 1 - v = x \qquad \text{or} \qquad v = 1 - x.$$

Therefore

$$x^2 = 1 - x.$$

Since m is the positive root of this equation, we have shown that the ratio of the side of a regular pentagon to the diagonal is m. Q.E.D.

5.10. Relatives of the Pentagon

The Greeks got two bonuses for studying the pentagon. First the dodecahedron and (somewhat more subtly) the icosahedron use the pentagon, and also the golden rectangle. This will be seen in Figures 5.13 and 5.14, but no further discussion is offered since this would take us into solid geometry. (The details are not difficult and the reader is urged to consult the articles by Coxeter and Gardner previously mentioned.)

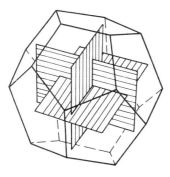

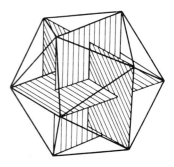

Figure 5.13. The regular dodecahedron (12 faces, 30 edges, 20 vertices)

Figure 5.14. Corners of three golden rectangles coincide with corners of a regular icosahedron (20 faces, 30 edges, 12 vertices)

Second the Greeks discovered the pentagram and thought it very attractive; see Figure 5.15. The Pythagoreans used it as some sort of club-membership pin and attached symbolic value to it. It is easy to see in it the suggestion of a perpetual movement. From a mathematical point of view it is extremely interesting as a visualization of a rotation of the plane which is of period five (repeating itself indefinitely), but is different from the rotation associated with the more common pentagon.

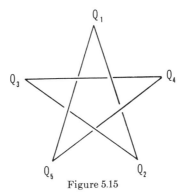

Figure 5.15

The reader will notice that the over-and-under passing in the figure is reminiscent of illustrations of clover-leaf intersections on modern high-speed highways. This over-and-under drawing is the modern way in mathematics of representing *knots*. If the pentagram is actually made out of string it will be found that the string is knotted. It is undoubtedly the purest coincidence that one can make a very good pentagon (in fact a precise pentagon, theoretically) by tying a knot in a strip of paper; the illustrations in Figure 5.16 show how. If this is done and held up to the light, there is seen the symbol of the pentagram.

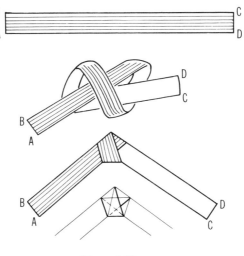

Figure 5.16

CHAPTER SIX

Constructions and Proofs

This final chapter covers much the same material as the previous pages of the book, but it is organized along different lines. The emphasis here is upon construction and proof; the word "construction" is used in a general sense that includes the definition of infinite processes. At the same time, the propositions we shall work out in this chapter answer some of the problems mentioned earlier.

6.1 Indirect Proof

There have already been several examples in the book of indirect proofs, also called proofs by contradiction. We showed through this device that $\sqrt{2}$ is irrational, also that the golden mean is irrational. Here we give a simpler sample of this most important technique.

ASSERTION. There does not exist any pair of integers (positive or negative) satisfying

$$4x + 6y = 15.$$

PROOF. Assume for the sake of argument that there is such a pair of integers; call them m and n. Then

$$15 = 4m + 6n = 2(2m + 3n).$$

This asserts that 15 is an even integer $(2m + 3n$ being the quotient on division by 2) which is false. This concludes the proof.

Problems

6.1. Accept the fact that $\sqrt{2}$ is irrational. Prove from this fact that the reciprocal of $(1 + \sqrt{2})$ is also irrational.

6.2. Prove that the form $ax + by$, where a and b are integers, represents an integer c through some choice of integers x and y if and only if c is divisible by the highest common factor of a and b. Thus, for example, the equation

$$12x + 18y = c$$

is satisfied by integers x and y only when c is divisible by 6.

6.2 A Theorem of Euclidean Geometry on Parallels as Proportional Dividers

HYPOTHESIS: t and t' are two lines (called transversals) which cut each of three parallel lines l_1, l_2, l_3, as shown in Figure 6.1.

CONCLUSION: The lengths of the line segments cut off from t and t' by the three parallel lines are proportional, that is,

$$\frac{AB}{BC} = \frac{A'B'}{B'C'}.$$

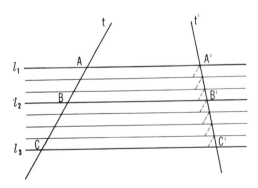

Figure 6.1

EXAMPLE. Inspection of the ruling in Figure 6.1 shows that

$$\frac{AB}{BC} = \frac{3}{4} \quad \text{and similarly} \quad \frac{A'B'}{B'C'} = \frac{3}{4},$$

confirming the theorem.

PROOF. The example also illustrates the proof in the general rational case, that is when AB/BC is a ratio of integers, say m/n.

98   USES OF INFINITY

a) *The rational or commensurable case.* In the rational case one makes l_1, l_2, l_3 part of a system of rulings which exhibit the ratio m/n. One divides AB into m equal intervals and BC into n equal intervals of the same length and constructs parallels at the points of division.

To prove that the ratio m/n is now transferred to t' one has to show that the segments on the right are all equal to each other.

For this purpose one constructs certain auxiliary segments, all parallel to t and each forming one side of a *parallelogram* when taken with the corresponding segment of t, and forming one side of a triangle when coupled with the corresponding segment on t'. It is now easy to prove that all these triangles are congruent; this rests on three types of elementary propositions: i) triangles are congruent if one side and all angles of one are equal to the corresponding parts in another; ii) opposite sides of a parallelogram are equal; iii) corresponding angles made by a line transversal to two parallels are equal. See Figure 6.2.

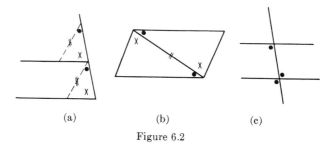

(a)   (b)   (c)

Figure 6.2

b) *The incommensurable case: Proof by continuity.* When AB/BC is not rational, the method just shown is not available. Let us now suppose, for example, that AB/BC is our familiar golden ratio $\frac{1}{2}(\sqrt{5} - 1)$. Let us take the sequence of *Fibonacci ratios* u_n defined by

$$u_n = \frac{f_{n-1}}{f_n} \qquad n = 1, 2, 3, \cdots,$$

where $f_{n+1} = f_n + f_{n-1}$, and the first terms of the sequence f_i are $f_1 = 1$, $f_2 = 1$, $f_3 = 2$, $f_4 = 3$, $f_5 = 5$, $\cdots$. It can be shown that, as n gets larger and larger, these Fibonacci ratios u_n approach the golden ratio m.† In symbols,

$$\lim_{n \to \infty} u_n = m.$$

† For a proof of this statement, see C. D. Olds, *Continued Fractions*, to appear in this series.

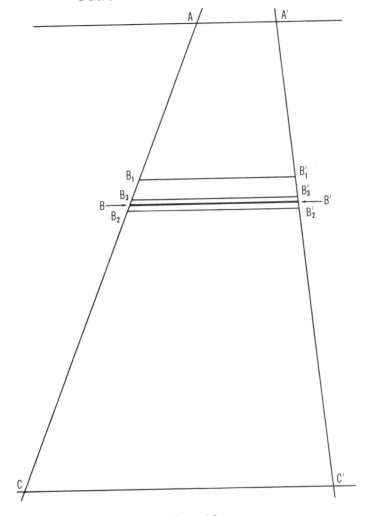

Figure 6.3

Now, letting the endpoints A and C of the segment AC stay fixed, we construct a sequence of points B_n on this segment such that for each n,

$$\frac{AB_n}{B_nC} = u_n ;$$

then the B_n approach B and the ratios u_n approach m. At each B_n we construct a parallel to BB' (see Figure 6.3), so obtaining a point B'_n on $A'C'$.

It follows from the rational case that

$$\frac{A'B'_n}{B'_n\,C'} = u_n\ ;$$

therefore, as n increases, these ratios approach m. What remains for us to prove is the intuitively obvious fact that the point B' is the limit of the constructed sequence $B'_1,\ B'_2,\ B'_3,\ \cdots$.

The reason for regarding this as an "intuitive fact" (in mathematics intuition is aways respected but not relied upon too heavily) is the true impression we get from Figure 6.3, namely that the variable point B'_n varies with the point B_n in such a way that the segments $A'B'_n$ approach the segment $A'B'$ *because* the segments AB_n approach the segment AB. Another way of saying this is that the points B'_n vary *continuously* with the points B_n. We shall show that, if the points B'_n approach a limit, this limit must be B'. Therefore, to prove the continuity, we require a geometric principle (not present in the geometry of the Pythagoreans) from which we can infer the existence of a limit. This principle is the geometric counterpart of the Bolzano-Weierstrass principle, discussed earlier in the book.

Using this principle we find that the sequence $B'_1,\ B'_2,\ B'_3,\ \cdots$ does have a limit point; call it B'' for the moment. B'' is a limit point of two sequences, $B'_1,\ B'_3,\ B'_5,\ \cdots$ lying on $A'B'$ and $B'_2,\ B'_4,\ B'_6,\ \cdots$ lying on $B'C'$. It follows now that B'' lies on both segments $A'B'$ and $B'C'$; but their only common point is B' and so $B'' = B'$. This concludes the proof of the theorem in case the ratio is given as m. But we have used no particular properties of m excepting only that it is a limit of a sequence of rational numbers which approach it from above and below. This is true of all real numbers, and therefore the proof is entirely general.

c) *The incommensurable case: Proof by contradiction.* The Greek geometers did not use the continuity proof just described but relied on an indirect argument. Assume that the assertion is false, that is, that

$$\text{either}\quad \frac{AB}{BC} > \frac{A'B'}{B'C'}\quad \text{or}\quad \frac{A'B'}{B'C'} > \frac{AB}{BC}.$$

We shall show that this assumption leads to a contradiction.
Let us assume first that

$$\frac{AB}{BC} > \frac{A'B'}{B'C'}\ ;$$

then there is a rational number $r = m/n$ (m and n are integers), such that

(6.1)
$$\frac{AB}{BC} > r > \frac{A'B'}{B'C'}.$$

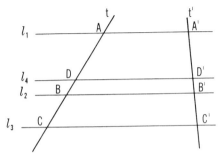

Figure 6.4

Let D be a point on t between A and C such that

$$\frac{AD}{DC} = \frac{m}{n}$$

(see Figure 6.4). Then D lies between A and B, for otherwise AB/BC would be less than or equal to r. Construct a line l_4 through D parallel to l_2; it intersects t' at D'. Now D' lies between A' and B' for otherwise the parallel segments BB' and DD' would intersect. Hence

$$\frac{A'B'}{B'C'} > \frac{A'D'}{D'C'}.$$

But by what we know about the rational case,

$$\frac{A'D'}{D'C'} = \frac{AD}{DC} = r, \qquad \text{so that} \qquad \frac{A'B'}{B'C'} > r,$$

which contradicts the inequality (6.1). Thus the assumption that

$$\frac{AB}{BC} > \frac{A'B'}{B'C'}$$

led to a contradiction and hence is false.

Similarly we can show that the assumption

$$\frac{A'B'}{B'C'} > \frac{AB}{BC}$$

leads to a contradiction. Therefore

$$\frac{AB}{BC} = \frac{A'B'}{B'C'}.$$

6.3 Definition by Recursion

A sequence of mathematical objects

first object, second object, third object, $\cdots$

is said to be constructed *recursively* whenever the description of each new object uses the fact that the previous one is already defined. For example, consider the sequence of points

(6.2) $$P_1, \quad P_2, \quad P_3, \quad \cdots$$

on the x-axis whose coordinates S_n satisfy the following conditions:

(6.3)
$$S_1 = 1,$$
$$S_n = S_{n-1} + \frac{1}{n}, \qquad n = 2, 3, 4, \cdots .$$

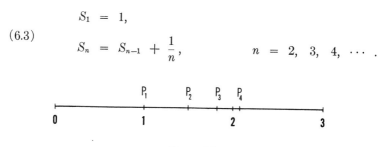

Figure 6.5

Note for contrast that the sequence of points

(6.4) $$Q_1, \quad Q_2, \quad Q_3, \quad \cdots$$

on the x-axis, defined by their x-coordinates a_n as

(6.5) $$a_n = n^2,$$

are independent of each other; that is, we can find Q_5 from the formula (6.5) without having found Q_4. To find P_5 we are directed by (6.3) to find P_4 first.

Let us suppose that some Zeno among our acquaintances has challenged us: "Does (6.3) define an *infinite* sequence?" He might say: "I grant that P_1 is defined, and P_2, and P_3, and as many more points as you have patience to construct, but no more. Thus (6.3) only shows how to construct a large finite sequence." Mathematicians do not hesitate to regard (6.3) as a definition of the sequence (6.2) for all values of n, i.e., $n = 1, 2, 3, \cdots$, just as they accept (6.5) as a definition of the sequence (6.4) for all values of n. The reader will recall that this recursive construction was used in the case of some of the later zig-zags in Chapter 4, and also in the spiralling rectangles and the zig-zag in Chapter 5 where, it is hoped, he accepted them as quite natural procedures.

To argue the matter a little more fully, one might paraphrase (6.3) as follows, in words. It consists of two parts. A *first step*, namely that the first number is 1. Next, a *continuing step*, which is the heart of the matter: *You are at a certain stage and have calculated a certain number. In order to calculate the next number, you merely add to this one the reciprocal of the total number of stages.*

Notice that this continuing instruction is not to be discarded after it has been used; on the contrary, it is worded so as to be always applicable to the next stage. Notice finally that this continuing instruction corresponds to the way in which the natural numbers are given to us: *First step*—the first number is 1. *Next step*—the next number is one more than the one you now have.

Thus (6.3) sets up a one-to-one correspondence between the points of the sequence (6.2) and the natural numbers. This is exactly what the formula (6.5) does for the sequence (6.4); (6.3) and (6.5) are equally valid ways of constructing a sequence.

We close this section with the mention of some interesting recursions. One of the most important recursively defined sequences in mathematics is this one:

$$1, \quad 2, \quad 6, \quad 24, \quad 120, \quad 720, \quad 5040, \quad 40320, \quad 362880, \quad 3628800, \quad \cdots .$$

This is the sequence designated by "$n!$" (read "n factorial") and defined thus:

$$1! = 1$$
$$(n + 1)! = (n + 1) \text{ times } n!, \qquad n = 1, \quad 2, \quad 3, \quad \cdots .$$

The concept of recursive definition embraces more general situations. For example, the Fibonacci series is recursively defined through the conditions that

$$f_1 = 1,$$
$$f_2 = 1;$$
$$f_{n+1} = f_n + f_{n-1}, \qquad n = 2, \quad 3, \quad 4, \quad \cdots .$$

Similarly, one can define a sequence recursively by such conditions as

$$b_1 = 1,$$
$$b_2 = 1 + 1 = 2,$$
$$b_3 = 1 + 1 + 2 = 4,$$
$$b_4 = 1 + 1 + 2 + 4 = 8;$$
$$b_n = 1 + b_1 + b_2 + \cdots + b_{n-1}, \qquad n = 2, \quad 3, \quad 4, \quad \cdots .$$

Problem

6.3. Can you identify this last sequence?

6.4 Induction

A formal proof by mathematical induction has two striking land-marks, a *first step* and an *inductive step*. Thus it is built exactly like a recursive construction and for the same reason: It is an infinite sequence of assertions arranged in an outwardly finite form. Let us consider a few examples and discuss the general principle later.

EXAMPLE 1. *Assertion:* The integer $1 + 4^n$ is not divisible by 3 for any value of $n = 1, 2, 3, \cdots$.

PROOF BY MATHEMATICAL INDUCTION. *First step*: The assertion is true for $n = 1$, because $1 + 4 = 5$ is not divisible by 3. The first step is proved.

Inductive step: IF the assertion is true for some integer k (that is, IF $1 + 4^k$ is not divisible by 3) THEN it is also true for the integer $k + 1$ (that is, $1 + 4^{k+1}$ is not divisible by 3). The proof of the inductive step follows. Notice that, since $4^{k+1} = 4 \cdot 4^k$,

$$(1 + 4^{k+1}) - (1 + 4^k) = 4^{k+1} - 4^k = 4^k \cdot (4 - 1) = 3 \cdot 4^k.$$

This shows that the difference between $(1 + 4^{k+1})$ and $(1 + 4^k)$ is divisible by 3; therefore one of them is divisible by 3 if and only if the other one is.† But since $1 + 4^k$ is *not* divisible by 3 (by the *hypothesis* of the inductive step), it follows that $1 + 4^{k+1}$ is not. This is the conclusion of the inductive step which is hereby proved.

Problems

6.4. Recall the definition: *An integer m has residue r modulo an integer q* means r *is the remainder when n is divided by q.* Prove the "Rule of 3": An integer N and the sum of its digits have the same residue modulo 3; for example, 47,158 and $25 = 4 + 7 + 1 + 5 + 8$, both have residue 1 modulo 3.

6.5. Prove by induction:

$$1 + 2 + \cdots + n = \frac{1}{2} n(n + 1), \qquad n = 1, 2, 3, \cdots.$$

---

† If $a - b = 3s$ (s is our integer), then $a = b + 3s$ and $b = a - 3s$. It follows that a is divisible by 3 if and only if b is, and that b is divisible by 3 if and only if a is.

CONSTRUCTIONS AND PROOFS

EXAMPLE 2. *Assertion* (as suggested by inspection of the multiplication table):

$$(1 + 2 + \cdots + n)^2 = 1^3 + 2^3 + \cdots + n^3, \quad n = 1, \ 2, \ 3, \ \cdots.$$

PROOF. *First step*: When $n = 1$, the left side is $1^2 = 1$ and the right side is $1^3 = 1$.

Inductive step: Suppose that k is such that

$$(6.6) \qquad (1 + \cdots + k)^2 = 1^3 + \cdots + k^3$$

and let us study

$$(6.7) \qquad (1 + \cdots + k + k + 1)^2.$$

This has the form $(A + B)^2$ with $A = 1 + 2 + \cdots + k$, and $B = k + 1$. Using $(A + B)^2 = A^2 + 2AB + B^2$, we get from (6.7) the following:

$$(1 + \cdots + k)^2 + 2(1 + \cdots + k)(k + 1) + (k + 1)^2.$$

If we apply to the middle term the formula to be proved in Problem 6.5 (with $n = k$), we get

$$(1 + \cdots + k)^2 + [k \cdot (k + 1)](k + 1) + (k + 1)^2.$$

Notice now that the last two terms have a common factor, namely $(k + 1)^2$. If we collect terms, this gives us

$$(6.8) \qquad (1 + \cdots + k)^2 + (k + 1)^3.$$

So far we have not used our hypothesis (6.6); if we now apply it to (6.8), this becomes

$$(6.9) \qquad 1 + 2^3 + \cdots + k^3 + (k + 1)^3.$$

In other words, the expression (6.7) has been transformed into (6.9). This concludes the proof of the inductive step, that IF our assertion holds for an integer k, THEN it holds for $k + 1$; the proof by induction is now complete.

The *logic* of these proofs is clear. We show that every one of an infinite sequence of statements is true, by proving that we can go through them one after another indefinitely and never find one that is false. We make sure that we have not missed any by starting with the first.

The startling thing is that all of this is done in what looks like a single (inductive) step but the fact is that this is a continuing step which is constantly reapplied by the mechanism of the "IF ⋯ THEN" proof; there is really an infinity of arguments, one for each case.

The *principle of the least integer* which we used for showing the irrationality of $\sqrt{2}$ and the *principle of infinite descent*, which we used to show the irrationality of m, the golden section, are closely related to proof by mathematical induction. It takes only a little practice to rephrase proofs so that they have one rather than the other form. However, we shall leave these considerations to the diligent reader.†

6.5 An Application to a Proof

The following is a valid proof of the fact that, for all n, the sum of the first n consecutive positive integers is $\frac{1}{2}n(n + 1)$. We mentioned this in Problem 6.5. The reader is urged to interpret the proof below as an instance of definition by recursion and proof by induction.

Let

$$S = \quad 1 \quad + \quad 2 \quad + \quad 3 \quad + \ldots + \quad n$$

Then S may also be written

$$S = \quad n \quad + n - 1 + n - 2 + \ldots + \quad 1$$

Adding these expressions, one obtains

$$2S = n + 1 + n + 1 + n + 1 + \ldots + n + 1 = n\,(n + 1).$$

Therefore

$$S = \tfrac{1}{2}n \cdot (n + 1),$$

and this holds for $n = 1,\ 2,\ 3,\ \cdots$. This argument is said to have been rediscovered by C. F. Gauss (1771–1855) when he was nine years old.

6.6 Fast Growing Sequences

The following story is a slightly modernized version of a calculation performed by Archimedes, combined with a several thousand years old tale for children.

Suppose that in the year 1750 some pair of mosquitos took cover in the state of New Jersey and suppose that they then generated a

---

† See the discussion of mathematical induction in Chapter 1 of *What Is Mathematics?* by Richard Courant and Herbert Robbins. New York: Oxford University Press, 1941.

population which has since doubled with each succeeding year. By 1760, there would have been 2^{10} descendants of this pair; this is about 10^3 and not very many. By 1770 they would have had a progeny of 2^{20} individuals which is over 10^6, or one million. By 1800 the number would be 2^{50} which is more than ten quadrillion (10^{16}). By now the population would number 2^{210}, which is more than 10^{63}.

To form some idea of what this number means (1 followed by 63 zeros is easier to say than to comprehend) let us suppose that one million small mosquitos will go into a box measuring one inch on a side. This gives us 10^{57} boxes. Out of these boxes, with careful stacking, we can make a giant cube of boxes measuring 10^{19} boxes on a side, and of course each side would be 10^{19} inches long. That is more than 10^{14} miles long. The distance from the sun to the earth is only about 90,000,000 miles or less than 10^8 miles, and Pluto is at a distance less than $4 \cdot 10^{10}$. So, if the sun were set in the middle of such a box even Pluto would be lost deep inside of it; indeed there would be room in the box for more than

$$(10^3)^3 = 10^9 = \text{one billion of our solar systems.}$$

Let us think of the numbers 1, 2, 3, 4, $\cdots$ and also the numbers 2, 4, 8, 16, 32, 64, $\cdots$ as representing the successive census-records of some growing populations; then the first sequence grows larger and larger without limit, but the second grows "out of this world" faster than the first.

Let us give this idea a mathematical form, and then see what we can prove. Table 6.1 shows the two sequences, and also the ratios of corresponding terms (except that I have rounded off these ratios to a whole number by dropping any fractional part).

TABLE 6.1

n	1	2	3	4	5	6	7	8	9	10	11
2^n	2	4	8	16	32	64	128	256	512	1024	2048
$\dfrac{2^n}{n}$	2	2	2	4	6	10	18	32	56	100	186

Table 6.2 indicates that 2^n is always greater than n, that from the 4th term on, 2^n is greater than n^2, and that from the 10th term on, 2^n is greater than n^3.

TABLE 6.2

n	1	2	3	4	5	6	7	8	9	10	11
2^n	2	4	8	16	32	64	128	256	512	1024	2048
n^2	1	4	9	16	25	36	49	64	81	100	121
n^3	1	8	27	64	125	216	343	512	729	1000	1331

At this point mathematical induction has acquired a new wrinkle. We can say that for all n (greater than zero, that is) 2^n exceeds n, but we cannot say that for all n, 2^n exceeds n^2. We may conjecture the following theorems.

THEOREM 1: *For all n greater than zero, 2^n exceeds n.*

THEOREM 2: *For all n greater than 4, 2^n exceeds n^2.*

THEOREM 3: *For all n greater than 9, 2^n exceeds n^3.*

Before we go on with more conjectures, let us talk about proofs. The problem before us is this: Can one use mathematical induction, even if one wants to start it not at $n = 1$, but at $n = 4$, or $n = 10$, or even later places?

The answer is yes. The reason is simple. *Any statement that is made concerning the sequence $N + 1$, $N + 2$, $N + 3$, $\cdots$ can immediately be translated into a statement about the sequence $N + n$ for all integers n.* This concludes the argument.

To illustrate the argument, take Theorem 2 and rewrite it like this: *For all integers n, 2^{n+4} exceeds $(n + 4)^2$.* Do you agree that if Theorem 2 is true, this new statement is true *for all n?*

Problem

6.6. Reformulate Theorem 3 in accordance with this program.

The proof of Theorem 1 is very simple. The first step consists of checking the statement for $k = 1$, and it is true.

Now for the inductive step: Let k denote an integer for which 2^k exceeds k. Then 2^{k+1} exceeds $2k$, and since $2k$ is at least as big as $k + 1$ it follows that 2^{k+1} exceeds $k + 1$, which concludes the proof.

The proof of Theorem 2 is only a little harder; it uses the fact that $(n + 1)^2 = n^2 + 2n + 1$. The proof is interesting, as you will see, because the inductive step and the first step are out of step.

The first step is at $k = 5$, and it is true that 2^5 exceeds 5^2.

For the inductive step, suppose that k is an integer for which 2^k exceeds k^2. Then 2^{k+1} exceeds $2k^2$. Now, when k exceeds 4, k^2 exceeds $4k$, and $2k^2$ exceeds $k^2 + 4k$ which exceeds $k^2 + 2k + 2$ which exceeds $(k + 1)^2$ *and we are through.*

Problem

6.7. Show that the inductive step in Theorem 2 is valid when k exceeds 2. Explain why this fact does *not* enable you to prove that 2^n exceeds n^2 for all n greater than 2.

The proof of Theorem 3 is slightly harder because we need the following case of the binomial theorem:

$$(n + 1)^3 = n^3 + 3n^2 + 3n + 1.$$

From this it is easy to conclude that $2n^3$ exceeds $(n + 1)^3$ provided that n exceeds 3. After that we can prove the inductive step in Theorem 3 when n exceeds 3, BUT the *first* step is false when $n = 4$, and we must start with the case n exceeds 9.

Although we have shown how to prove Theorem 3, we will do it over again in the standard algebraic symbols used for this proof. The word "exceeds" is replaced by the symbol $>$.

First step of Theorem 3: For $k = 10$, $2^{10} > 10^3$.
Inductive step: If $2^k > k^3$, and $k > 9$ then

$$2^{k+1} > 2k^3 = k^3 + k^3 > k^3 + 9k^2 > k^3 + 3k^2 + 6k$$
$$> k^3 + 3k^2 + 3k + 3$$
$$> k^3 + 3k^2 + 3k + 1$$
$$= (k + 1)^3.$$

This concludes the proof of the inductive step and of the theorem.

Theorems 4, 5, 6, $\cdots$ can be guessed but are harder to prove because they require more knowledge of the binomial theorem. However, we ought to talk about them, in a general way at least, because IF we wanted to prove an infinite sequence of theorems, depending on an integer, we would need to do this by mathematical induction.

This gives us a case of induction in which the first step is a theorem (2^n exceeds n) which is proved by induction.

The following theorem is true:

THEOREM N: *For every integer N there is a least integer K (depending on N) such that for all integers n greater than K*

$$2^n \text{ exceeds } n^N.$$

For $N = 1$, this becomes our Theorem 1, and K is the integer zero; for $N = 2$, we get Theorem 2, and K is 4; for $N = 3$, we get Theorem 3 and $K = 9$. It is reasonable to guess that for $N = 4, 5, 6, \cdots, 10$, $K = 16, 25, 36, \cdots, 100$ and that in general $K = N^2$.

The reader is urged to postpone the following set of problems until he has read through all of this section, because Theorem N has been introduced for two different purposes. First, it suggests how an infinite process (in this case mathematical induction) can be "cascaded", each stage becoming an infinite process; secondly, it provides an introduction to certain arguments by Cantor, one of the great masters of the "science of the infinite". This second purpose is more important than the first.

Problems

6.8. Corresponding to the step to determine the K of Theorem N for any integer N:

(a) Prove that the Nth power of the Nth power of 2 is the N^2th power of 2, i.e.,
$$(2^N)^N = 2^{N^2}.$$

(b) When N exceeds 4, 2^N exceeds N^2. Prove that then
$$2^{N^2} \text{ exceeds } (N^2)^N,$$
and hence that if $n = N^2$, then $2^n > n^N$.

6.9. Corresponding to the inductive step in Theorem N: Prove that 2^n exceeds n^N provided only that n exceeds N^2.

6.10. From the preceding problems, show by induction on N, starting with $N = 1$, that Theorem N holds.

Theorem N is very important, and I should like to call attention to it by numerical instances. It says that 2^n increases faster than n^{10} (which it doesn't overtake until about the hundredth term) and that 2^n increases faster than n^{1000}, even though it doesn't catch up to that sequence until the millionth term. It is clear that the *powers* of n, n^2, n^3, n^4, $\cdots$ increase faster than n (with increasing n). Because 2^n increases faster than any power of n, it is said to increase *transcendentally* faster than n. Let us look at some other "high-speed sequences".

The sequence given by 4^n (with increasing n) increases faster than 2^n, but *not* transcendentally faster. Thus, for every n,

$$4^n = (2^2)^n = 2^{2n} = (2^n)^2.$$

Similarly, the sequences 5^n, 6^n, $\cdots$, N^n, for all integers N, are seen to increase like powers of the fast sequence 2^n. However there is a sequence which increases transcendentally faster than 2^n, as you can anticipate.

THEOREM. *Given any sequence (of positive increasing integers) a, b, c, d, $\cdots$, the sequence $2^{(2^a)}$, $2^{(2^b)}$, $2^{(2^c)}$, $\cdots$ increases transcendentally faster than the sequence 2^n.*

Thus the sequence 2^2, $2^{(2^2)}$, $2^{(2^3)}$, $\cdots$ increases transcendentally faster than the sequence 2, 2^2, 2^3, $\cdots$.

The theorem follows (by a bit of reasoning) as an application of Theorem N, used for all N.

The fact that there is always a transcendentally faster sequence than any given one suggests the following inductive construction. Suppose we take a fast sequence, like 2^n, then a transcendentally faster sequence, then a third, transcendentally faster than the second, and so on forever. Now where are we? Have we got the fastest "thing" that can be constructed?

The following theorem says no. The proof uses the Cantor diagonal-argument, which he invented in 1870, and which is one of the great ideas in the foundations of mathematics and has many applications in all branches of mathematics. We already used it in Chapter 3.

THEOREM. *Suppose that we are given a sequence S_1, S_2, S_3, $\cdots$ of sequences, each of which (after the first) increases faster than the preceding one. Let us arrange the entire system of numbers in a rectangular array, the numbers S_k occupying the kth row, the nth term of S_k being in the nth column of the array. Then the diagonal sequence, namely the first term of the first sequence, followed by the second term of the second sequence, followed by the third term of the third sequence, and so on, increases faster than any one of the given sequences.*

To illustrate this theorem, let S_k denote the sequence of kth powers of n, i.e., n^k for $n = 1,\ 2,\ \cdots$. The first few terms of the first few rows of the rectangular array are displayed in Table 6.3.

TABLE 6.3

S_1	1	2	3	4	5	6	7	8	9	10	·
S_2	1	4	9	16	25	36	49	64	81	100	·
S_3	1	8	27	64	125	216	343	·	·	·	·
S_4	1	16	81	256	625	1296	·	·	·	·	·
S_5	1	32	243	1024	3125	·	·	·	·	·	·
S_6	1	64	729	·	·	·	·	·	·	·	·
⋮	⋮	⋮	⋮	⋮	⋮	⋮	⋮	⋮	⋮	⋮	⋮

The *diagonal sequence* is:

$$1,\quad 4,\quad 27,\quad 256,\quad 3125,\quad \cdots\ ;$$

it is easy to see that the general term is n^n. This sequence increases transcendentally faster than n, and in fact it also happens to increase transcendentally faster than 2^n.

6.7 Dirichlet's Boxes or the Pigeon Hole Principle

There is an important technique of proof known as the *principle of Dirichlet's boxes*. It says that if one has more objects than boxes into which to put them, and one puts all the objects into the boxes, then at least one box will have to hold more than one object. A regular application of this occurs in the problem of newspaper reporters and hotel rooms during important national conventions.

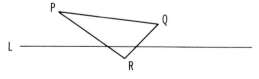

Figure 6.6

a) *The finite case.* A simple application of this principle is as follows: Let L denote a line in the plane and let P, Q, and R be three distinct points no one of which is on the line; then at least one of the segments PQ, QR, or RP does not meet the line. This is because a line divides the plane into only two "sides", and, of the three points, two must lie on the same side and can be joined by a segment not meeting the line. The assertion follows.

b) *A countable infinity of objects, a finite number of boxes. The Fibonacci sequence.* Next we shall apply Dirichlet's box principle to an interesting theorem on the Fibonacci sequence. But first we must recall some facts about divisibility properties and residues.

Let N be a given integer. If any other integer is divided by N, the following remainders are possible: 0 (if the integer is divisible by N), 1, 2, 3, $\cdots$, $N-1$. In other words, the remainder (also called *residue, modulo N*) is always one of these N numbers. In particular, if we picked more than N numbers, say

$$b_1, \quad b_2, \quad \cdots, \quad b_N, \quad b_{N+1}, \quad \cdots \quad b_m,$$

and divided them all by N, we know from the box principle that at least two of our b's must have the same residue.

We now ask the following question: Suppose we have a sequence

$$b_1, \quad b_2, \quad b_3, \quad \cdots$$

of whole numbers and we write down the sequence of their residues modulo N,

$$r_1, \quad r_2, \quad r_3, \quad \cdots.$$

Will *two consecutive remainders* ever be repeated in this sequence? In other words, is there a pair r_i, r_{i+1} and another pair, r_j, r_{j+1}, such that

$$r_i = r_j \quad \text{and} \quad r_{i+1} = r_{j+1}\,?$$

To answer this question, let us first figure out how many different *pairs* of numbers can be formed from the possible remainders

$$0, \quad 1, \quad 2, \quad \cdots, N-1.$$

Since the first number of the pair can have any one of N values, and since the second can also have any one of N values, there are N^2 distinct ordered pairs possible. Now a sequence of $t+1$ terms

$$b_1, \quad b_2, \quad b_3, \quad \cdots, \quad b_t, \quad b_{t+1}$$

has the t consecutive pairs

$$(b_1, b_2), \quad (b_2, b_3), \quad \cdots, \quad (b_{t-1}, b_t), \quad (b_t, b_{t+1}).$$

Therefore, by the box principle, the residues (mod N) of any sequence of more than N^2+1 numbers must contain at least two identical pairs of consecutive terms.

EXAMPLE. If $N = 3$, the possible values for residues are 0, 1 and 2. All the possible $3^2 = 9$ pairs then are

0, 0;   0, 1;   0, 2;   1, 0;   1, 1;   1, 2;   2, 0;   2, 1;   2, 2.

Consider the sequence of $11 = N^2 + 2 > N^2 + 1$ consecutive subway stations on the 8th Avenue Independent Line:

4,   14,   23,   34,   42,   50,   59,   72,   81,   86,   96.

The sequence of residues (mod 3) is

1,   2,   2,   1,   0,   2,   2,   0,   0,   2,   0;

since we have written down $3^2 + 2 = 11$ stops, we know that in the sequence of residues some pair is repeated. In fact, the second and third terms and the sixth and seventh constitute identical pairs; so do the fifth and sixth, and the ninth and tenth.

We could now ask: after how many terms of a sequence would we get a repeated consecutive triplet of residues (mod N), or a repeated consecutive foursome? These questions can be answered by the same reasoning, but we shall stick to pairs and apply our knowledge to a special sequence, the Fibonacci sequence

1,   1,   2,   3,   5,   8,   13,   21,   34,   55.

(6.10)   89,   144,   233,   377,   610,   987,   1597,

2584,   4181,   6765,   10946,   17711, $\cdots$

which is defined by the recursion relation

$$f_1 = f_2 = 1,$$

(6.11)

$$f_{n+1} = f_n + f_{n-1} \qquad \text{for } n \geq 2.$$

We shall prove the following observation due to Lagrange:†

THEOREM. *Let* N *be any integer greater than or equal to 2. Then the residues* (mod N) *of the Fibonacci sequence repeat after at most* $N^2 + 1$ *terms. More precisely, the very first pair of residues, namely* 1,   1, *occurs again within* $N^2 + 2$ *terms and from then on the entire sequence is repeated.*

† This observation can be found in Euler's textbook on elementary algebra (written in 1766 by dictation, soon after he became blind). It was translated into many languages, and Lagrange himself did the French version.

Before giving the proof, let us test the assertion in a few simple cases.

For $N = 2$, the residues of (6.10) are

$$1, \quad 1, \quad 0, \quad 1, \quad 1, \quad 0, \quad 1, \quad 1, \quad 0, \quad 1, \quad 1, \quad 0, \quad \cdots.$$

The fourth and fifth terms equal the first and second, and the same three terms, 1, 1, 0, seem to repeat over and over. Note that $N^2 + 1 = 5$ and it is within 5 terms that a pair is repeated.

For $N = 3$, the residues of (6.10) are

$$1, \quad 1, \quad 2, \quad 0, \quad 2, \quad 2, \quad 1, \quad 0, \quad 1, \quad 1, \quad 2, \quad 0, \quad 2, \quad 2, \quad 1, \quad 0, \quad \cdots.$$

The 9th and 10th terms are the same as the 1st and 2nd. Here

$$N^2 + 1 = 10.$$

For $N = 4$, the residues of (6.10) are

$$1, \quad 1, \quad 2, \quad 3, \quad 1, \quad 0, \quad 1, \quad 1, \quad 2, \quad 3, \quad 1, \quad 0, \quad \cdots;$$

the 7th and 8th terms begin the repetition. Note that $N^2 + 1 = 17$; from the box principle we know that a repeated pair has to occur within 18 terms, but it actually occurs much sooner.

For $N = 5$, the residues of (6.10) are

$$1, \quad 1, \quad 2, \quad 3, \quad 0, \quad 3, \quad 1, \quad 4, \quad 0, \quad 4, \quad 4,$$
$$3, \quad 2, \quad 0, \quad 2, \quad 2, \quad 4, \quad 1, \quad 0, \quad 1, \quad 1, \quad 2, \quad \cdots;$$

here $N^2 + 1 = 26$, but the first pair occurs already within 21 terms, and the cycle seems to repeat.

PROOF OF LAGRANGE'S THEOREM. In addition to the box principle (which told us that the sequence of residues of any old sequence will exhibit a repeating consecutive pair within $N^2 + 2$ terms), we must make use of the special feature of the Fibonacci sequence

$$f_1, \quad f_2, \quad \cdots,$$

that it is defined by the recursion formula (6.11). This relation tells us that, once a pair of residues is repeated, i.e., once we find a pair f_k, f_{k+1} which has the same residues as an earlier pair f_i, f_{i+1}, then we must expect that f_{k+2} and f_{i+2} have the same residues, that f_{k+3} and f_{i+3} have the same residues, etc.

To see this, recall that "f_i and f_k have the same residue (mod N)" means they have the same remainder upon division by N, i.e.,

$$f_i = q_i N + r_i, \qquad f_k = q_k N + r_k, \qquad r_i = r_k ;$$

or, written in another way,

$$f_i \equiv r_i \;(\text{mod } N), \qquad f_k \equiv r_k \;(\text{mod } N), \qquad r_i = r_k ;$$

so that

$$f_i \equiv f_k \;(\text{mod } N).$$

The symbols $u \equiv v \;(\text{mod } N)$, read "$u$ is congruent to v modulo N", simply mean that $u - v$ is divisible by N. Note that, *if*

$$u \equiv v \;(\text{mod } N) \qquad and \qquad x \equiv y \;(\text{mod } N),$$

then

$$u + x \equiv v + y \;(\text{mod } N) \qquad and \qquad u - x \equiv v - y \;(\text{mod } N),$$

because if $u - v$ and $x - y$ are divisible by N, then so is their sum $(u - v) + (x - y) = (u + x) - (v + y)$, and so is their difference $(u - v) - (x - y) = (u - x) - (v - y)$.

In this notation the result from the box principle says that there are terms in the sequence for which

$$f_i \equiv f_k \;(\text{mod } N) \qquad and \qquad f_{i+1} \equiv f_{k+1} \;(\text{mod } N).$$

It follows that

$$f_i + f_{i+1} \equiv f_k + f_{k+1} \;(\text{mod } N).$$

But from the recursion relation (6.11), we see that the left member of this congruence is just f_{i+2} and the right member is f_{k+2}. Therefore

$$f_{i+2} \equiv f_{k+2} \;(\text{mod } N).$$

The same reasoning shows that

$$f_{i+3} \equiv f_{k+3} \;(\text{mod } N),$$

$$f_{i+4} \equiv f_{k+4} \;(\text{mod } N),$$

$$\cdots\cdots\cdots\cdots\cdots\cdots$$

So far we have shown that the sequence of residues (mod N) of the Fibonacci sequence is periodic, but we must still show that the cycle begins at the beginning, with the pair 1, 1.

This is easy. Suppose the *period* of the sequence is p, so that

$$(6.12) \qquad\qquad f_j \equiv f_{j+p} \pmod{N} \qquad\qquad (j \geq s)$$

for all j from a certain one on, say $j = s$. We want to show that f_s is in fact the first term of the sequence.

If $s > 1$, then f_s is not the first term in the sequence, but then we can find the previous term, f_{s-1}, from the recursion formula, and argue as follows:

$$\begin{aligned} f_{s-1} &= f_{s+1} - f_s \\ &\equiv f_{s+1+p} - f_{s+p} \pmod{N} \qquad\qquad \text{by (6.12)} \\ &= f_{s+p-1} \qquad\qquad \text{by the recursion relation.} \end{aligned}$$

Hence,

$$f_{s-1} \equiv f_{s-1+p} \pmod{N}.$$

If $s - 1 = 1$, the repeating cycle begins with f_1 and the proof is complete.

If $s - 1 > 1$, we show that

$$f_{s-2} \equiv f_{s-2+p} \pmod{N}$$

by the same reasoning as above. Since s is some finite integer, this process applied successively to $s - 1$, $s - 2$, $s - 3$, $\cdots$ will eventually be applied to 1 and will lead to

$$f_1 \equiv f_{1+p} \pmod{N},$$

and this completes the proof of Lagrange's theorem.

Observe that the proof made use only of the recursion formula

$$f_{n+1} = f_n + f_{n-1}, \qquad\qquad n \geq 2,$$

and not of the *initial conditions*, $f_1 = f_2 = 1$. This means that Lagrange's theorem is valid for any sequence satisfying the relation $a_{n+1} = a_n + a_{n-1}$ for $n \geq 2$, no matter what values we take for a_1 and a_2.

Problems

6.11. Consider the sequence 1, 1, 5, 13, 41, $\cdots$ built on the rule that, for $n > 2$, $a_{n+1} = 2a_n + 3a_{n-1}$. Prove an analogue to the Lagrange remark, but notice that modulo 6 the sequence runs 1, 1, 5, 1, 5, $\cdots$.

6.12. Generalize Lagrange's remark to all sequences built as follows: for $n > 2$, $a_{n+1} = \alpha a_n + \beta a_{n-1}$, where α and β are arbitrary given integers.

6.13. Generalize to the case that after the third term,

$$a_{n+1} = \alpha a_n + \beta a_{n-1} + \gamma a_{n-2} .$$

6.14. Generalize still further.

c) *An uncountable infinity of objects, a countable infinity of boxes.*
A use of Dirichlet's boxes in the case of infinite sets is illustrated by
the following. Suppose that somehow there has been selected for us
in the plane an uncountable infinity of rectangles. The remarkable
thing about the theorem we are going to prove is that it does not
matter how these rectangles are chosen—all that matters is that
there shall be an uncountable infinity of them.

It is asserted that there exists *some* circle and there exists *some
uncountable* set of the given rectangles such that the circle is inside
every one of the rectangles. That is to say, it lies in the interior and
does not meet the rectangles' perimeters. The proof follows; it is
technically quite easy, but may appear quite difficult on a first
meeting.

We prove that we can select the circle from among the "rational
circles", that is, circles which have radii of rational lengths and whose
centers have rational coordinates. The set of these is countable since

$$\aleph_0^3 = \aleph_0 ;$$

cf. Section 3.9. The proof is by contradiction. Thus, if no one of the
rational circles is inside uncountably many rectangles of the some-
how-given set of rectangles, then each and every one of them is inside
of a countable infinity of our rectangles *at most* (possibly a finite
number, possibly none). Now it follows, since $\aleph_0 \cdot \aleph_0 = \aleph_0$ (cf.
Section 3.9) that the totality of rectangles of our set, each of which
has a rational circle inside of it, is at most a countable set of rectangles. Since the given set is uncountable, there must be rectangles
left over; that is, *there must be rectangles in our given set which do not
have any rational circles inside of them!*

This remark has been emphasized because it is absurd. *Every*
rectangle has at least one rational circle inside of it (infinitely many,
as is easily seen), and we have been led to a contradiction completing
the proof.   The statement relied on above, that *every* rectangle has
rational circles inside of it, is true. The assertion that this is easily
seen is in conformity with mathematical tradition which attacks one
difficulty at a time and is optimistic about later ones. The proof of
this assertion follows.

Let a rectangle be given in the plane and let P denote its center;
see Figure 6.7. If both coordinates of P are rational, then choose
P as center of a circle and choose as radius any rational number

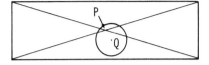

Figure 6.7

which is smaller than one half the shorter side. However, if the coordinates of P are not both rational then (as is easily done) choose a point Q with both coordinates rational which is inside the rectangle (P is a limit point of a sequence of such points). Now find the distances from Q to each one of the sides and choose a rational number smaller than all four of these distances as radius, using Q as center. This completes the proof that every rectangle has rational circles inside of it.

What we have proved illustrates the important distinction in analysis between countable and uncountable sets. Notice how easy it is to construct a countable set of rectangles which do not overlap each other; we have shown that every uncountable set must overlap very substantially.

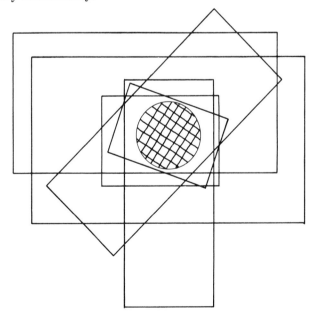

Figure 6.8. An instance of overlapping rectangles where the substantial overlap is measured by a circle contained in every one of them

The following problems are proved along the lines shown by the preceding discussion. The reader will find them worth thinking about, but they are not easy. The second problem is meant to serve as one of the details for the third.

Problems

6.15. For the moment, let us call a rectangle *special* if it is related to the coordinate system as follows: its sides lie on the lines

$$x = a, \quad y = b, \quad x = c, \quad y = d,$$

with $a \neq c$ and $b \neq d$ and a, b, c, d rational. Prove that the set of these rectangles is countably infinite.

6.16. Let P be a point inside a given rectangle. Show that then P is inside of a special rectangle which is inside of the given one.

6.17. Suppose that you are given an uncountable set of points in the plane called X. Prove that there exists a point P contained in X such that every rectangle containing P contains uncountably many points of X. (The point P is called an accumulation point of the set X.)

6.18. Suppose you are somehow given an uncountable set X of points on a line (or in the plane). Prove that there exists a sequence of points P_1, P_2, P_3, $\cdots$ in X and a point P, also in X, such that P is the sequential limit point of the sequence P_1, P_2, P_3, $\cdots$.

This is difficult to show because one has so little knowledge about X. It is introduced here because it is a powerful tool, like the Bolzano-Weierstrass principle of the least upper bound, for finding limits. It represents one of the most important uses of an uncountable infinity.

Solutions to Problems

CHAPTER TWO

2.1 This is a non-terminating sequence of sets of musical compositions, the first set consisting of compositions for one voice part or instrument, the second set of pieces for two performers, the third of pieces for three performers, and so on.†

2.2 This is a periodic sequence of the four classes of hits in baseball. The iteration dots indicate that we are to repeat the same sequence of classes again and again.

2.3 Collections of siblings born on the same day make up the terms of this sequence. The first term is the collection of all individuals with one such sibling, the second is the set of individuals with two such siblings, etc. The terms which occur beyond a certain point in this infinite sequence are empty sets.

2.4 Here we have a list of the names of the days of the week. In this case the iteration dots represent an abbreviation for the days Saturday and Sunday.

2.5 This is a periodic sequence, the terms of which are the first letters in the names of the days of the week, in the order of the days, beginning with the letter M corresponding to Monday. The first term occurs again after six more terms, and from then on the entire period is repeated over and over.

2.6 The first of these three sets is the collection of all integers n of the form $n = 3q$ where q is an integer. It is easily seen that this set consists of the numbers 0, 3, 6, 9, $\cdots$.

The second set is composed of all integers n of the form $n = 1 + 3q$. Since each such number has the remainder 1 when divided by 3, the numbers of the infinite sequence 1, 4, 7, 10, $\cdots$ belong to the second set.

---

† Remark added by the author: I can see "quintet", "sextet", "septet", "octet", but there I get stuck. In the absence of a clear-cut rule as to just how to continue, I would agree with a student who called the question unclear and would count all answers correct.

The numbers in the third set are each of the form $2 + 3q$, where q is an integer, and hence each has remainder 2 when divided by 3. Thus, the third set has the elements 2, 5, 8, 11, $\cdots$.

Inasmuch as every integer when divided by 3 has one and only one of the remainders 0, 1, 2, we know that these three nonoverlapping infinite sets together comprise the entire set of integers.

2.7 If q divides n, then $n = qb$ and $n + 1 = qb + 1$, where b is an integer. In other words, if n is divisible by q, $n + 1$ has the remainder 1 when divided by q, and hence n and $n + 1$ have no common factor.

2.8 The general principle which is suggested by an examination of these tables is that for every k by k multiplication table, where k is any positive whole number, the sum of the numbers in the table is the square of the sum of the first k positive integers. The sum of the integers in any lower gnomon-figure is the cube of the smallest integer in that gnomon.

2.9 $\dfrac{1}{7} = 1 \cdot 10^{-1} + 4 \cdot 10^{-2} + 2 \cdot 10^{-3} + 8 \cdot 10^{-4} + 5 \cdot 10^{-5} + 7 \cdot 10^{-6} + \dfrac{1}{7} \cdot 10^{-7}$

$= .142857 + .000000142857 + \dfrac{1}{7} \cdot 10^{-14} = .142857142857 \cdots$.

Similarly, $\dfrac{1}{9} = 1 \cdot 10^{-1} + \dfrac{1}{9} \cdot 10^{-2} = .111111 \cdots$.

$\dfrac{1}{11} = 0 \cdot 10^{-1} + 9 \cdot 10^{-2} + \dfrac{1}{11} \cdot 10^{-3} = .090909 \cdots$.

$\dfrac{1}{99} = 0 \cdot 10^{-1} + 1 \cdot 10^{-2} + \dfrac{1}{99} \cdot 10^{-3} = .010101 \cdots$.

Every terminating decimal may be written in the form $a/10^k$ where a is an integer; for example,

$$3.572 = \frac{3572}{10^3}.$$

If a contains factors 2^s and/or 5^t with $0 < s$, $t \le k$, then $a/10^k$ is not in lowest terms and may be reduced as follows:

$$\frac{a}{10^k} = \frac{2^s \cdot 5^t \cdot b}{2^k \cdot 5^k} = \frac{b}{2^{k-s} \cdot 5^{k-t}} = \frac{b}{2^n \cdot 5^m}.$$

2.10 The symbols $0.9090909090 \cdots$ and $0.0909090909 \cdots$ are the non-terminating decimal representations of $\frac{10}{11}$ and $\frac{1}{11}$, respectively. When we "add" these expressions the resulting symbol is $.9999999999 \cdots$, whereas the sum of $\frac{10}{11}$ and $\frac{1}{11}$ is $\frac{11}{11} = 1$.

2.11 Since

$$n \div m = n \times \frac{1}{m},$$

the method is clear. For example, to divide 7 by 9 we look up the reciprocal of 9 in the table of Figure 2.4 and write

$$7 \div 9 = 7 \times \frac{1}{9} = 7 \times .111 \cdots = .777 \cdots.$$

2.12 Construct the circle of radius PQ with center Q; see Figure 2.7. Using P as center, and the same radius, swing the compass to find the points P_1 and Q_1 of intersection of this circle with the first circle.

Next, open the compass to the width P_1Q_1; draw circles with center P_1 and Q_1 respectively. One of their intersections (the one to the right of P and Q) is the desired point R.

To see that P, Q, R are collinear and $PQ = QR = d$, observe that P_1PQ and Q_1PQ are two equilateral triangles with common base PQ which is bisected by the segment P_1Q_1 in the point M (see Figure 2.7), and that P_1Q_1R is an equilateral triangle with base P_1Q_1 and whose altitude MR lies on the line through P and Q.

Moreover, if $PQ = d$, then

$$P_1Q_1 \;=\; 2P_1M \;=\; d\sqrt{3} \qquad \text{and} \qquad MR \;=\; \frac{1}{2}P_1Q_1\sqrt{3} \;=\; \frac{3d}{2} \;;$$

$$QR \;=\; MR \;-\; MQ \;=\; \frac{3d}{2} \;-\; \frac{d}{2} \;=\; d.$$

This method shows us how to construct an infinite sequence of points on a line. Simply pick two points, call them P and Q, get R, and repeat the above construction on Q and R, getting R', etc.

2.13 Suppose P is to the left of Q. Line up the ruler with P and Q so that its right end is at Q and make a mark on the ruler at the place where P falls. This mark divides the ruler into two parts, one of length $PQ = d$ on the right, and one of smaller length d' on the left. After drawing the segment PQ, line up the ruler with the segment PQ so that the division mark on the ruler falls on Q. Then the right end of the ruler will be at a distance d from Q, on PQ extended. Denote the endpoint of this extension by Q_1. Next move the ruler in the direction from Q to Q_1 along the line until the division mark is on the new point Q_1. The right endpoint of the ruler will be at a point Q_2 on the line through P and Q, at a distance $2d$ from Q. Continue this process indefinitely in order to extend the line through P and Q indefinitely to the right.

In order to extend the line through P and Q to the left, we just reflect the method just described.

If d' were longer than d, the same method would work but it would be more economical to interchange the roles of d and d'.

2.14 Once the direction of the road and the point at which it is to enter the mountain are determined, it is only necessary to line up every three consecutive guide-posts. This can be checked at each advance.

2.15 The successive midpoints approach the point which is at a distance from A equal to $\frac{2}{3}$ of the length of AB.

2.20 (a) Divide a segment into 5 equal parts, hold one part, and give 3 pieces away. Then one part remains and we hold $\frac{1}{4}$ of the amount which has been distributed. If we repeat the same process over and over, each time dividing the one remaining part into 5 equal pieces, we shall continue to hold $\frac{1}{4}$ of the total amount distributed and a smaller and smaller amount of the original segment will remain to be distributed.

Thus

$$\frac{1}{5} + \frac{1}{25} + \frac{1}{125} + \cdots = \frac{1}{4}.$$

Similarly,

(b)
$$\frac{1}{8} + \frac{1}{64} + \frac{1}{512} + \cdots = \frac{1}{7};$$

(c)
$$\frac{1}{10} + \frac{1}{100} + \frac{1}{1000} + \cdots = \frac{1}{9}.$$

2.21 Divide the segment into n equal parts, hold m of them, leave m of them to work on and give away the remaining $n - 2m$ parts. Thus $m + (n - 2m) = n - m$ parts have been distributed, and we hold m parts; hence we hold $m/(n - m)$ of the distributed amount.

We treat the remaining part in the same way, dividing it into n equal pieces, holding m, giving away $n - 2m$ and keeping m to be worked on. Continuing in this way, we shall always hold $m/(n - m)$ of the distributed part while the part to be worked on gets smaller and smaller.

2.22 It follows from $4 = 2 \cdot 2$ that $\sqrt{4} = 2$, so that $\sqrt{4}$ is rational. To prove that $\sqrt{3}$ and $\sqrt{5}$ are irrational, we need only consider the following:

(a) If we assume that $\sqrt{3} = p/q$, where p/q is that fraction (among all equivalent fractions) which has the smallest denominator, then we have

$$1 < \frac{p}{q} < 2,$$
$$q < p < 2q,$$
$$0 < p - q < q,$$

and

$$3q^2 = p^2,$$
$$3q^2 - pq = p^2 - pq,$$
$$q(3q - p) = p(p - q),$$

which implies, in contradiction to our initial assumption, that

$$\frac{p}{q} = \frac{3q - p}{p - q}.$$

(b) Assume that $p/q = \sqrt{5}$, p/q the fraction with smallest denominator. Then $p = \sqrt{5}q$ and $2 < p/q < 3$ give us

$$2q < p < 3q,$$
$$0 < p - 2q < q,$$

and

$$p^2 = 5q^2,$$

$$p^2 - 2pq = 5q^2 - 2pq,$$

$$p(p - 2q) = q(5q - 2p),$$

so that

$$\frac{p}{q} = \frac{5q - 2p}{p - 2q},$$

where $p - 2q < q$.

2.23 From $p = \sqrt{7}q$ and $2 < p/q < 3$ we get

$$2q < p < 3q,$$

$$0 < p - 2q < q;$$

$$p^2 = 7q^2,$$

$$p^2 - 2pq = 7q^2 - 2pq,$$

$$p(p - 2q) = q(7q - 2p).$$

Then

$$\frac{p}{q} = \frac{7q - 2p}{p - 2q}$$

contradicts the assumption that $\sqrt{7}$ can be represented as a ratio of two integers, p and q, where q is the smallest possible denominator.

2.24 $\sqrt{8} = \sqrt{4\cdot2} = \sqrt{4}\cdot\sqrt{2} = 2\sqrt{2}$. If $\sqrt{8} = p/q$, then $2\sqrt{2} = p/q$, or $\sqrt{2} = p/2q$ (a contradiction since $\sqrt{2}$ is irrational).

2.25 $p = \sqrt{n}q$ and $k^2 < n < (k + 1)^2$ imply that

$$k^2 < \left(\frac{p}{q}\right)^2 < (k + 1)^2,$$

$$k < \frac{p}{q} < k + 1,$$

$$kq < p < (k + 1)q,$$

$$0 < p - kq < q.$$

Since $p^2 = nq^2$,

$$p^2 - kpq = nq^2 - kpq,$$

$$p(p - kq) = q(nq - kp),$$

$$\frac{p}{q} = \frac{nq - kp}{p - kq} \quad \text{where } p - kq < q.$$

If there were a fraction whose square is the integer N, we would write it with as small a denominator as possible, say $p/q = \sqrt{N}$, and $q \neq 1$ by assumption, so $q^2 \neq 1$. Hence the fraction p^2/q^2 would lie between consecutive integers k and $k + 1$ and we can produce the above contradiction.

2.26 If $\sqrt{s}$ is not an integer, then, by problem 2.25, s is not the square of a fraction. But if $2\sqrt{s} = n$, where n is an integer, then s is equal to the fraction $(n/2)^2$. Hence, unless $\sqrt{s}$ is an integer, we are involved in a contradiction.

2.27 To prove that $12.5 \times a = 100a/8$ we need only note that

$$12.5 = 12 + \frac{1}{2} = \frac{24 + 1}{2} = \frac{25}{2} = \frac{100}{8}.$$

2.28 Let a be any integer with digits $a_k a_{k-1} \cdots a_0$. Then we may write

$a = a_k 10^k + a_{k-1}10^{k-1} + \cdots + a_1 \cdot 10 + a_0$

$= a_k(10^k - 1 + 1) + a_{k-1}(10^{k-1} - 1 + 1) + \cdots + a_1(10 - 1 + 1) + a_0$

$= a_k 99 \cdots 9 + a_k + a_{k-1} 99 \cdots 9 + a_{k-1} + \cdots + a_1 \cdot 9 + a_1 + a_0$

$= 3[a_k(33 \cdots 3) + a_{k-1}(33 \cdots 3) + \cdots + a_1 \cdot 3]$

$\qquad\qquad\qquad\qquad + a_k + a_{k-1} + a_{k-2} + \cdots + a_1 + a_0.$

Since the first expression on the right is a multiple of 3, a has the same remainder upon division by 3 as the second term on the right, i.e. the sum of the digits of a. Another way of saying this is that a and the sum of its digits belong to the same residue class (mod 3).

If $a_k + a_{k-1} + \cdots + a_0 < 10$, the proof is complete. If not, let

$a_k + a_{k-1} + \cdots + a_0 = b = b_j 10^j + b_{j-1}10^{j-1} + \cdots + b_1 \cdot 10 + b_0,$

and proceed in the same manner as above; eventually the sum of digits will be less than 10, and we shall have reached the root number $r(a)$. This shows that a and the sum b of its digits and the sum c of the digits of b etc. down to $r(a)$ are all in the same residue class modulo 3.

2.29 (a) and (b) Let us consider the infinite sequence of decimals

$$a_1 = .1111\cdots,$$

$$a_2 = .101010\cdots,$$

$$a_3 = .100100100\cdots,$$

$$a_4 = .1000100010001\cdots,$$

$$\cdots\cdots\cdots\cdots\cdots\cdots,$$

$$a_k = .1000\cdots1000\cdots1000\cdots,$$

and note that the period of a_k is k.

(c) $.1010010001000010000010000001\cdots$.

CHAPTER THREE

3.1 The proof of this theorem for the case $AB':B'B = m:n$, where m and
     n are positive integers, is essentially the same as the proof given in
     Chapter 6, Section 6.2(a). For $m = \sqrt{2}$, $n = 1$, the theorem can be
     proved by the methods used in Sections 6.2(b) and 6.2(c) for the incom-
     mensurable case.

3.2 Such a rectangle does not exist because it would lead to the equation
     $1 = 0 \cdot x$, where x is the length of the other side. But this contradicts
     the rule of arithmetic: $x \cdot 0 = 0$ for all x whatsoever.

3.3 Construct the right triangle ADC (see Figure 3.6) with legs of lengths x
     and 1. Draw the perpendicular to the hypotenuse AC through C. Ex-
     tend line AD to meet this perpendicular at B. The segment DB then
     has the desired length $y = 1/x$ because the length of the altitude CD of
     right triangle ACB is the mean proportional between the lengths of
     $AD = x$ and $DB = y$:

$$\frac{y}{1} = \frac{1}{x}.$$

3.4 The parabola $y = x^2$ is the locus of points (x, y) such that, for every
     abscissa x, the ordinate y satisfies the relation

$$\frac{y}{x} = \frac{x}{1}.$$

This relation suggests again the construction of a right triangle so that
the altitude to the hypotenuse has length x and divides the hypotenuse
into segments of lengths 1 and y. In Figure 3.7 this construction was
carried out for two given values of x, x_1 and x_2, and the corresponding values of y are y_1 and y_2. (The details of this construction are left
to the reader.)

The parabola shown in Figure 3.8 may now be plotted either by using
values (x, y) obtained from the construction of Figure 3.7 as coordinates
for points of the graph, or it may be plotted directly by carrying out the
construction in the coordinate plane as indicated: For each abscissa x,
find the point $Z: (x, -1)$, connect it to the origin O, draw a perpendicular
to the resulting segment at O and locate the point $W: (x, y)$ at which this
perpendicular intersects the vertical line x units away from the y-axis.
In the right triangle OWZ the altitude to the hypotenuse has length x
and divides ZW into segments of length 1 and y so that the relation

$$\frac{1}{x} = \frac{x}{y} \quad \text{or} \quad y = x^2$$

is satisfied for each W so constructed.

3.5 If x is a given length then $\sqrt{x}$ can always be constructed by virtue of the relation

$$\frac{1}{\sqrt{x}} = \frac{\sqrt{x}}{x}.$$

For example, in Figure 3.7, let $AE = 1$ as before, extend AE to B so that EB has the given length x, and draw the semicircle with AB as diameter. The perpendicular to AB at E will intersect this semicircle at a point O, and EO, the altitude of right triangle AOB, will have length $\sqrt{x}$.

The length $\sqrt{x}$ can also be read off the parabola of Figure 3.8: just find a point whose distance from the horizontal axis is x; its distance from the vertical axis is $\sqrt{x}$.

3.6 To find approximately the cube root $\sqrt[3]{a}$ of a number a one determines the point of intersection P of the graph of $y = x^3$ with the horizontal line $y = a$ and measures the abscissa of P. The coordinates of P are $(\sqrt[3]{a}, a)$.

3.8 A careful construction will show that as n gets larger and larger, the quantity $\sqrt{n + 1} - \sqrt{n}$ becomes smaller and smaller.

3.9 Let $a = \sqrt{n + 1}$, $b = \sqrt{n}$. Then

$$(\sqrt{n + 1} + \sqrt{n}) \cdot (\sqrt{n + 1} - \sqrt{n}) = 1,$$

or

$$\sqrt{n + 1} - \sqrt{n} = \frac{1}{\sqrt{n + 1} + \sqrt{n}}.$$

Clearly, as n gets larger and larger the denominator $\sqrt{n + 1} + \sqrt{n}$ increases, and it follows from the above identity that the quantity $\sqrt{n + 1} - \sqrt{n}$ becomes smaller and smaller; that is, the difference between $\sqrt{n + 1}$ and $\sqrt{n}$ can be made as small as we wish (it is always greater than 0) by taking n large enough.

3.10 We multiply the expression by

$$\frac{\sqrt{2n + 1} + \sqrt{2n}}{\sqrt{2n + 1} + \sqrt{2n}}$$

and obtain

$$\sqrt{n} \; \frac{2n + 1 - 2n}{\sqrt{2n + 1} + \sqrt{2n}} = \frac{\sqrt{n}}{\sqrt{2n + 1} + \sqrt{2n}} = \frac{1}{\sqrt{2 + \dfrac{1}{n}} + \sqrt{2}}.$$

As n gets larger and larger, $1/n$ becomes smaller and smaller so that this expression approaches

$$\frac{1}{2\sqrt{2}} = \frac{\sqrt{2}}{4}.$$

3.11 If $0 < y < \frac{1}{2}\pi$, then   (see Figure 3.15), $\sin y < y < \tan y$; dividing by $\tan y$, we obtain

$$\cos y < \frac{y}{\tan y} < 1.$$

As y decreases, $\cos y$ approaches 1 so that $y/\tan y$ is squeezed between 1 and a number close to 1.

3.12 (a) PROOF OF THEOREM 3.1. Let S be the sequence x_1, x_2, x_3, $\cdots$ having the limit L, and suppose that y_1, y_2, y_3, $\cdots$ is any subsequence S' of S (i.e., S' is an infinite sequence whose terms are some or all of the terms of S, arranged in the order in which they occur in S).

Now S has the limit L means that the sequence $(x_1 - L)$, $(x_2 - L)$, $(x_3 - L)$, $\cdots$ approaches 0; that is, for every integer n, there are at most a finite number (depending on n) of terms $x_k - L$, $k = 1,$ 2, 3, $\cdots$, which are numerically larger than $1/n$. But if S' is a subsequence of S, then every term of the sequence $(y_1 - L)$, $(y_2 - L)$, $\cdots$ is identical to some term $x_k - L$, so that, for every integer n, there can be at most a finite number of terms $y_{k'} - L$, $k' = 1,$ 2, 3, $\cdots$, numerically larger than $1/n$.

Thus, every subsequence of an infinite sequence with limit L also has the limit L.

(b) PROOF OF THEOREM 3.2. $a + x - (a + L) = x - L$, so that if $(x_1 - L)$, $(x_2 - L)$, $(x_3 - L)$, $\cdots$ approaches zero, then $[a + x_1 - (a + L)]$, $[a + x_2 - (a + L)]$, $\cdots$ also approaches zero; therefore $a + x_1$, $a + x_2$, $a + x_3$, $\cdots$ has the limit $a + L$.

(c) PROOF OF THEOREM 3.3. We wish to show that, for every integer m, all but a finite number of terms kx_i satisfy

$$|kx_i - kL| < \frac{1}{m}$$

provided that the sequence x_i has L as limit. In other words we know that for every n, all but a finite number of x_i satisfy

$$|x_i - L| < \frac{1}{n}.$$

In particular, take n to be the nearest integer greater than or equal to $|k|m$. Then

$$|kx_i - kL| = |k||x_i - L| < |k|\frac{1}{n} \le |k|\frac{1}{|k|m} = \frac{1}{m}$$

for all but a finite number of terms. This argument holds for every integer m.

(d) We prove first: if x_1, x_2, x_3, $\cdots$ has the limit L and y_1, y_2, y_3, $\cdots$ has the limit M, then x_1y_1, x_2y_2, x_3y_3, $\cdots$ has the limit LM.

Following the hint, we write

$$x_i y_i - LM = x_i y_i - x_i M + x_i M - LM$$
$$= x_i (y_i - M) + M(x_i - L).$$

By assumption, for every integer n.

$$|x_i - L| < \frac{1}{n} \quad \text{and} \quad |y_i - M| < \frac{1}{n}$$

for all but a finite number of x_i, y_i. Moreover, since the x_i have a limit, all but a finite number are certainly bounded by some constant, say C. Thus

$$|x_i y_i - LM| \leq |x_i||y_i - M| + M|x_i - L|$$
$$\leq C \frac{1}{n} + M \frac{1}{n} = (C + M) \frac{1}{n}.$$

Now, given any integer m, we can achieve

$$|x_i y_i - LM| \leq \frac{1}{m}$$

merely by choosing the integer n so that

$$\frac{C + M}{n} < \frac{1}{m},$$

thus establishing, for every integer m,

$$|x_i y_i - LM| < \frac{1}{m}$$

for all but a finite number of terms $x_i y_i$.

To prove Theorem 3.4, that the limit of x_1^2, x_2^2, $\cdots$ is L^2, use the above result with $y_i = x_i$ and $L = M$. To prove that the limit of x_1^k, x_2^k, $\cdots$ is L^k, we simply repeat the argument $k - 1$ times.

3.13 (a) This problem is not at all easy. Take the case $0 < r < 1$. One standard proof runs as follows:

Since $1/r > 1$, $1/r = 1 + h$, $h > 0$. Now

$$\left(\frac{1}{r}\right)^n = (1 + h)^n > 1 + nh$$

for every n, and so $(1 + h)^n$ increases without limit as n increases. Therefore the reciprocal, r^n, goes to zero.

The assertion that $(1 + h)^n$ increases without limit as $n \to \infty$ can also be proved by appeal to the Bolzano-Weierstrass principle described in the next section. For if L is any number such that

$$(1 + h)^n \leq L$$

for all n, then

$$(1 + h)^{n-1} \leq \frac{L}{1 + h} = L' < L$$

for all n; and so, given *any* upper bound L to this sequence, there would exist a *smaller* upper bound L'. Such a sequence is incompatible with the Bolzano-Weierstrass principle.

Let us suppose now that $r = 1$. Then $r^2 = 1$, $r^3 = 1$, $r^4 = 1$, $\cdots$ and each number of the sequence $(r - 1)$, $(r^2 - 1)$, $\cdots$ equals 0. Since no term of this sequence exceeds $1/n$, $n > 0$, the sequence approaches 0 and r, r^2, r^3, $\cdots$ has the limit 1 when $r = 1$.

(b) From the identity

$$a(1 - x)(1 + x + x^2 + \cdots + x^{n-1}) = a(1 - x^n)$$

we get

$$a \cdot 1 + ar + ar^2 + \cdots + ar^{n-1} = \frac{a(1 - r^n)}{1 - r};$$

then, since r^n approaches 0 if $|r| < 1$, we have

$$\lim_{n \to \infty} \frac{1 - r^n}{1 - r} = \frac{1}{1 - r}, \qquad -1 < r < 1.$$

Hence, the sum of any infinite geometric series $a + ar + ar^2 + \cdots$ with ratio r numerically smaller than 1 is $a/(1 - r)$.

3.14 The sequence of rational numbers constructed in the text, in decimal form, is

$$1.\,0\,000000000\cdots,$$
$$.5\,0\,00000000\cdots,$$
$$2.00\,0\,0000000\cdots,$$
$$.333\,3\,333333\cdots,$$
$$3.0000\,0\,00000\cdots,$$
$$.25000\,0\,0000\cdots,$$
$$.666666\,6\,666\cdots,$$
$$1.5000000\,0\,00\cdots,$$
$$4.00000000\,0\,0\cdots,$$
$$.200000000\,0\cdots,$$
$$\cdots\cdots\cdots\cdots.$$

We shall construct a number whose kth decimal is always one or the other of the digits 2 or 3 but differs from the kth decimal of the kth number in this list.

Let us choose 3 as its first decimal since 3 is different from the first decimal of the first number. Let us choose 3 for the second decimal of the number we are constructing because 3 is different from the second decimal of the second number in the list. Similarly, let us choose $3 (\neq 0)$ for the third decimal, $2 (\neq 3)$ for the fourth decimal, $3 (\neq 0)$ for the fifth, $3 (\neq 0)$ for the sixth, $3 (\neq 6)$ for the seventh, etc., until we come to a kth number with 3 in the kth place; then we will choose 2.

3.15 Before classifying the letters of the alphabet, we shall consider, for the sake of definiteness, the letter T. We shall simplify the situation by considering an uncountable set of identical (i.e. congruent) letters T (e.g. with a stem of one inch and a top bar of one inch) whose vertices are labelled A, B, C, see Figure 3.18(a).

On any piece of paper of finite dimensions, an uncountable set of such T's must contain an uncountable subset of T's such that the vertices labelled A are within $\frac{1}{4}$, say, inch of each other. In this set, there is an uncountable subset such that the vertices labelled B are within $\frac{1}{8}$ inch of each other. And in this set, there is an uncountable subset such that the vertices labelled C are within $\frac{1}{8}$ inch of each other.

Now then, let ABC and $A'B'C'$ [see Figure 3.18(b)] be two T's of the kind described. They cross. This can be proved by elementary geometry from the fact that a straight line divides the plane into two regions (called the two *sides* of the line), and segments connecting points on opposite sides of the line must cross the line. This establishes the impossibility of writing an uncountable number of congruent non-crossing T's on a page. Note that if we merely required the distances AA' and BB' to be small, then the T's could possibly stand as they do in Figure 3.19, and not cross. We shall not treat the case of T's of varying sizes here, but the same result can be proved.

All letters of the alphabet that contain a configuration such as we encountered in the letter T (i.e. an intersection of two segments or curves where at least one of the segments extends beyond the point of intersection) are in the same class as T. They are A, B, E, F, H, K, P, Q, R, T, X, Y.

For all other letters, it is possible to scribble an uncountable set of them on a page. Figures 3.20(a), (b) illustrate this fact for the letters L, O respectively. In each case, the fact that a line segment contains an uncountable number of points gives the clue.

CHAPTER FOUR

4.1 The phrase "if this limit exists" has been omitted. The statement $L = \lim_{n \to \infty} L_n$ makes sense only if the L_n have a limit, and in this case it asserts that L is the value of this limit.

4.2 The direct computation of the lengths S_1, S_2, $\cdots$ of the sides of equilateral triangles whose bases are on the x-axis and whose vertices lie on the curve $y = x^2$ is somewhat awkward; fortunately the question posed in the problem can easily be answered without such a computation: The length of the resulting zig-zag is again 2 because, as in Example 2, it is twice as long as the distance from $(1, 0)$ to the origin.

4.3 We calculate the distances

$$B_1 T = 1, \quad B_3 T = \frac{1}{2}, \quad B_5 T = \frac{1}{4}, \quad \cdots, \quad B_{2n+1} T = \frac{1}{2^n}, \quad \cdots$$

which approach zero. Hence T is the limit of the sequence

$$B_1, \quad B_3, \quad \cdots \quad B_{2n+1}, \quad \cdots.$$

The distances $B_{2n}B_{2n-1}$ can be represented as sides of equilateral triangles of lengths $1/2^{n-1}$ and so these distances also approach 0.

By virtue of the triangle inequality, we have the following relations between lengths:

$$B_{2n}T \leq B_{2n}B_{2n-1} + B_{2n-1}T.$$

As n increases each term on the right approaches zero (by what we showed above) and hence their sum approaches zero. Therefore T is also the limit of points with even subscripts.

4.4 The point Z: $(\sqrt{2}, \sqrt{2}/3)$ is the limit of the sequence of points D_1, D_2, $\cdots$. The abscissas of D_1, D_2, $\cdots$ are

$$x_1 = \frac{\sqrt{2}}{2}, \quad x_2 = \frac{\sqrt{2}}{2} + \frac{\sqrt{2}}{4}, \quad x_3 = \frac{\sqrt{2}}{2} + \frac{\sqrt{2}}{4} + \frac{\sqrt{2}}{8}, \quad \cdots,$$

$$x_n = \frac{\sqrt{2}}{2}\left(1 + \frac{1}{2} + \cdots + \frac{1}{2^{n-1}}\right), \quad \cdots$$

When these finite geometric progressions are summed, they have the form

$$x_1 = \sqrt{2}\left(1 - \frac{1}{2}\right), \quad x_2 = \sqrt{2}\left(1 - \frac{1}{4}\right),$$

$$x_3 = \sqrt{2}\left(1 - \frac{1}{8}\right), \quad \cdots, \quad x_n = \sqrt{2}\left[1 - \left(\frac{1}{2}\right)^n\right], \quad \cdots;$$

these numbers come arbitrarily close to $\sqrt{2}$ since the sequence $\frac{1}{2}, \frac{1}{4}, \frac{1}{8}, \cdots$ approaches zero. This shows that $\lim_{n\to\infty} x_n = \sqrt{2}$, and this is the meaning of the phrase "the abscissa of Z is the limit of the abscissas of D_1, D_2, $\cdots$".

To prove that the ordinates

$$y_1 = \frac{\sqrt{2}}{2}, \quad y_2 = \frac{\sqrt{2}}{2}\left(1 - \frac{1}{2}\right), \quad y_3 = \frac{\sqrt{2}}{2}\left(1 - \frac{1}{2} + \frac{1}{4}\right),$$

$$\cdots, \quad y_n = \frac{\sqrt{2}}{2}\left[1 - \frac{1}{2} + \frac{1}{4} - \cdots + \left(-\frac{1}{2}\right)^{n-1}\right], \quad \cdots$$

of D_1, D_2, $\cdots$ have the limit $\sqrt{2}/3$, we sum these finite geometric series and find that

$$y_n = \frac{\sqrt{2}}{2}\left[\frac{1 - (-\frac{1}{2})^n}{1 + \frac{1}{2}}\right] = \frac{\sqrt{2}}{3}\left[1 - \left(-\frac{1}{2}\right)^n\right].$$

As $n \to \infty$, $(-\frac{1}{2})^n$ approaches zero so that the y_n have the limit $\sqrt{2}/3$.

From the fact that the abscissas have the limit $\sqrt{2}$ and the ordinates have the limit $\sqrt{2}/3$, we can prove that the sequence D_1, D_2, $\cdots$ has the limit Z: $(\sqrt{2}, \sqrt{2}/3)$ by the Pythagorean Theorem. We express the distance D_nZ by

$$(D_nZ)^2 = (\sqrt{2} - x_n)^2 + \left(\frac{\sqrt{2}}{3} - y_n\right)^2.$$

The terms on the right approach zero as $n \to \infty$, hence their squares approach zero, and so does the sum of their squares. Therefore the distances D_nZ approach zero and Z is indeed the limit of the sequence D_1, D_2, $\cdots$.

4.5 Denote the abscissas and ordinates of $D_1', D_2', \cdots$ by $x_1, x_2, \cdots$ and $y_1, y_2, \cdots$ respectively. It is clear from the construction of Example $4'$ and our knowledge of Example 4 that

$$x_1 = \frac{\sqrt{2}}{2}, \quad x_2 = \frac{\sqrt{2}}{2}\left(1 + \frac{1}{2}\right), \quad x_3 = \frac{\sqrt{2}}{2}\left(1 + \frac{1}{2} - \frac{1}{4}\right), \quad \cdots,$$

$$x_n = \frac{\sqrt{2}}{2}\left(1 + \frac{1}{2} - \frac{1}{4} - \frac{1}{8} + \cdots \pm \frac{1}{2^{n-1}}\right)$$

$$= \frac{\sqrt{2}}{2}\left[\left(1 - \frac{1}{4} + \frac{1}{16} - \frac{1}{64} + \cdots \pm \frac{1}{2^{n-2}}\right)\right.$$

$$\left. + \frac{1}{2}\left(1 - \frac{1}{4} + \frac{1}{16} - \frac{1}{64} + \cdots \pm \frac{1}{2^{n-2}}\right)\right]$$

$$= \frac{\sqrt{2}}{2}\left[\frac{1 - (-\frac{1}{4})^{n/2}}{\frac{5}{4}} + \frac{1 - (-\frac{1}{4})^{n/2}}{2 \cdot (\frac{5}{4})}\right]$$

$$= \frac{3\sqrt{2}}{5}\left[1 - \left(-\frac{1}{4}\right)^{n/2}\right], \qquad \text{for } n \geq 2, n \text{ even,}$$

and

$$x_{n+1} = x_n \pm \frac{\sqrt{2}}{2}\left(\frac{1}{4}\right)^{n/2}, \quad \text{so that} \quad \lim_{n \to \infty} x_n = \frac{3\sqrt{2}}{5},$$

and

$$\lim_{n \to \infty} x_{n+1} = \lim_{n \to \infty} x_n \pm \lim_{n \to \infty} \frac{\sqrt{2}}{2}\left(\frac{1}{4}\right)^{n/2} = \frac{3\sqrt{2}}{5}.$$

For the ordinates, we have

$$y_1 = \frac{\sqrt{2}}{2}, \quad y_2 = \frac{\sqrt{2}}{2}\left(1 - \frac{1}{2}\right), \quad y_3 = \frac{\sqrt{2}}{2}\left(1 - \frac{1}{2} - \frac{1}{4}\right), \quad \cdots,$$

$$y_n = \frac{\sqrt{2}}{2}\left(1 - \frac{1}{2} - \frac{1}{4} + \frac{1}{8} + \frac{1}{16} - \cdots \pm \frac{1}{2^{n-1}}\right)$$

$$= \frac{\sqrt{2}}{2}\left[\left(1 - \frac{1}{4} + \frac{1}{16} - \cdots \pm \frac{1}{2^{n-2}}\right)\right.$$

$$\left. - \frac{1}{2}\left(1 - \frac{1}{4} + \frac{1}{16} - \cdots \pm \frac{1}{2^{n-2}}\right)\right]$$

$$= \frac{\sqrt{2}}{2}\left[\frac{1 - (-\frac{1}{4})^{n/2}}{\frac{5}{4}} - \frac{1 - (-\frac{1}{4})^{n/2}}{2 \cdot (\frac{5}{4})}\right]$$

$$= \frac{\sqrt{2}}{5}\left[1 - \left(-\frac{1}{4}\right)^{n/2}\right], \qquad \text{for } n \geq 2, n \text{ even,}$$

and

$$y_{n+1} = y_n \pm \frac{\sqrt{2}}{2} \left(\frac{1}{4}\right)^{n/2},$$

so that

$$\lim_{n\to\infty} y_n = \frac{\sqrt{2}}{5} = \lim_{n\to\infty} y_{n+1}.$$

As we have seen in the solution to Problem 4.4, this implies that the point $(3\sqrt{2}/5, \sqrt{2}/5)$ is the limit point of the sequence D_1', D_2', $\cdots$.

4.6 Project each point E_1, E_2, E_3, and so on, perpendicularly onto the y-axis (this is what we do, essentially, when we calculate the ordinate of a point). Call these points F_1, F_2, F_3, and so on. Now if we let f_n denote the ordinate of E_n, then f_n is also the length of OF_n. Notice that the sequence of numbers f_2, f_4, f_6, $\cdots$ is constantly increasing and is bounded above. By the Bolzano-Weierstrass principle this sequence has a limit f^*. We shall show in a moment, but the reader may prefer to prove it himself, that f^* is also the limit of the sequence of odd-numbered f's, f_1, f_3, f_5, $\cdots$, which approach it from above. Thus the entire sequence has f^* as a limit, but the approach to this limit is two-sided. To f^* there corresponds a point F^* (on the y-axis) which ought to be the projection of the limit of the points E_1, E_2, E_3, $\cdots$; but as we have seen, the points E_1, E_2, $\cdots$ have no limit.

A proof that the sequence of odd-numbered f's has f^* as a limit follows. For every n,

$$f_{2n-1} - f^* = (f_{2n-1} - f_{2n}) + (f_{2n} - f^*),$$

the second term in parentheses being negative (see Figure 4.10). The first term on the right is precisely $\sqrt{2}/(2n)$. We have no formula for the second term; *but* since the sequence f_{2n} ($n = 1, 2, 3, \cdots$) converges to f^*, we know that all such terms are small when n is large enough, by the very definition of limit. Thus we can be sure that when n is large enough, the right hand side is the difference of two small numbers and is small. This shows that f_{2n-1} is near to f^* (for large n) and concludes the proof.

4.7 Example 5 showed (see solution to previous problem) that a sequence of points in the plane may have no limit point although the sequence of their projections has a limit point. Assertion (a) would be correct if one added "if the given sequence of points has a limit"; we shall demonstrate this in a moment.

Assertion (b) is true. Let Q_n be the points in the given sequence, Q its limit point, P_n the projections of Q_n, and P the projection of Q. Then $P_nP = Q_nQ \cos \alpha_n$ where α_n is the angle between the segment Q_nQ and the line which carries the projections. Since $|\cos \alpha_n| \leq 1$ it follows that $|P_nP| \leq |Q_nQ|$ and since $\lim_{n\to\infty} Q_nQ = 0$, it follows that $\lim_{n\to\infty} |P_nP| = 0$ so that P, the projection of Q, is indeed the limit point of the sequence P_n.

This also proves that the limit of the projections is the projection of the limit of a sequence of points, provided they have a limit. If P^* is the limit of the projections, P the projection of the limit Q, then $\lim_{n \to \infty} P_n P^* = 0$ and $\lim_{n \to \infty} P_n P = 0$ imply that P and P^* coincide.

4.8   Let

$$S_n = 1 + \frac{1}{2^2} + \frac{1}{3^2} + \cdots + \frac{1}{n^2}.$$

Since for $k > 1$ we have $k > k - 1$, it follows that

$$\frac{1}{k} < \frac{1}{k - 1} \quad \text{and} \quad \frac{1}{k^2} < \frac{1}{k(k - 1)} \qquad \text{for all } k > 1.$$

Therefore

(1) $$S_n < 1 + \frac{1}{1 \cdot 2} + \frac{1}{2 \cdot 3} + \cdots + \frac{1}{(n - 1)n}.$$

Now we observe that

$$\frac{1}{(k - 1)k} = \frac{1}{k - 1} - \frac{1}{k} \qquad \text{for } k = 2, 3, \cdots$$

and re-write the right member of (1) in the form

$$S_n < 1 + \left(\frac{1}{1} - \frac{1}{2}\right) + \left(\frac{1}{2} - \frac{1}{3}\right) + \cdots + \left(\frac{1}{n - 1} - \frac{1}{n}\right)$$

$$= 2 - \frac{1}{n}.$$

This proves that $S_n < 2$, for $n = 1, 2, \cdots$.

4.9   By Pythagoras' theorem, we may express any segment OP_k in terms of the previous one:

(1) $$OP_k^2 = OP_{k-1}^2 + \frac{1}{(k - 1)^2}.$$

Next, we express OP_{k-1} in terms of the previous one and substitute in (1):

$$OP_k^2 = OP_{k-2}^2 + \frac{1}{(k - 2)^2} + \frac{1}{(k - 1)^2}.$$

Continuing in this manner we find that

$$OP_k^2 = OP_1^2 + 1 + \frac{1}{2^2} + \frac{1}{3^2} + \cdots + \frac{1}{(k - 1)^2},$$

and, since $OP_1 = 1$, we have (in the notation used in the solution to the previous problem)

$$OP_k^2 = 1 + S_{k-1}.$$

We have seen that $S_n < 2$ for all n. Hence $OP_n^2 < 1 + 2 = 3$ and

(2) $$OP_n < \sqrt{3} \qquad\qquad \text{for all } n.$$

The length of the zig-zag $P_1 P_2 \cdots P_n$ is

$$L_n = 1 + \frac{1}{2} + \frac{1}{3} + \cdots + \frac{1}{n}$$

and we have already seen (page 65) that the harmonic series $1 + \frac{1}{2} + \frac{1}{3} + \cdots$ has no limit.

To prove that the projections Q_k (on a circle of radius 3 and center O) of our points P_k wind around indefinitely often, it suffices to show that the sum of the angles α_n between OP_n and OP_{n+1} becomes arbitrarily large. To see this, consider

$$\sin \alpha_n = \frac{\frac{1}{n}}{OP_{n+1}} = \frac{1}{n \, OP_{n+1}}.$$

By our result (2), we see that

$$\frac{1}{n \, OP_{n+1}} > \frac{1}{n\sqrt{3}}, \qquad \text{for } n = 1, 2, \cdots;$$

Moreover, for any acute angle α, we have $\sin \alpha < \alpha$ (see Figure 3.15). Thus

$$\alpha_1 + \alpha_2 + \cdots > \sin \alpha_1 + \sin \alpha_2 + \cdots > \frac{1}{\sqrt{3}} + \frac{1}{2\sqrt{3}} + \cdots$$

$$= \frac{1}{\sqrt{3}} \left(1 + \frac{1}{2} + \frac{1}{3} + \cdots \right),$$

so that the sum of the angles exceeds the harmonic series (multiplied by the constant factor $1/\sqrt{3}$) and hence is infinite.

The sequence $P_1, \quad P_2, \quad \cdots$ clearly cannot have a limit point for, if it did, all points after a certain point (say P_N) on would have to be in some small sector of the circle, and this is clearly not the case.

4.10 (a) $\qquad \frac{1}{2} + \frac{1}{4} + \frac{1}{6} + \cdots = \frac{1}{2} \left[1 + \frac{1}{2} + \frac{1}{3} + \cdots \right].$

The quantity in brackets is the harmonic series treated earlier. It was found to be infinite. Hence, a constant times the harmonic series is infinite, and the series (a) diverges.

(b) $\dfrac{1}{3} + \dfrac{1}{7} + \dfrac{1}{11} + \cdots = \dfrac{1}{4 \cdot 1 \, - \, 1} + \dfrac{1}{4 \cdot 2 \, - \, 1} + \dfrac{1}{4 \cdot 3 \, - \, 1}$

$$+ \cdots + \frac{1}{4n \, - \, 1} + \cdots$$

since

$$\frac{1}{4n \, - \, 1} \geq \frac{1}{4n} \qquad \text{for } n = 1, 2, \cdots,$$

each term of the series (b) is greater than the corresponding term in the diverging series

$$\frac{1}{4} + \frac{1}{8} + \frac{1}{12} + \cdots \frac{1}{4n} + \cdots = \frac{1}{4}\left[1 + \frac{1}{2} + \frac{1}{3} + \cdots\right],$$

and hence the series (b) diverges.

(c) Since for $n > 1$, $n > \sqrt{n}$, we have

$$\frac{1}{\sqrt{n}} > \frac{1}{n}$$

and

$$\frac{1}{1} + \frac{1}{\sqrt{2}} + \frac{1}{\sqrt{3}} + \cdots + \frac{1}{\sqrt{n}} + \cdots$$

$$> \frac{1}{1} + \frac{1}{2} + \frac{1}{3} + \cdots + \frac{1}{n} + \cdots.$$

Hence the sequence (c) diverges.

(d) The terms of this sequence are even larger than the corresponding terms of (c) and therefore (d) certainly diverges.

4.11 (a) If for every line in the plane the projection of P is the limit of the projections of P_n, then this is true, in particular, for the two perpendicular axes of a coordinate system. Denote the projections of P_n on the x-axis and on the y-axis by x_n, y_n respectively, and those of P by x and y. Then, see Figure 4.15(a),

$$(P_nP)^2 = (x_n - x)^2 + (y_n - y)^2,$$

and since the x_n approach x and the y_n approach y,

$$\lim_{n\to\infty} (P_nP)^2 = 0$$

and the P_n approach P.

(b) Clearly, this result cannot be deduced from the fact that the given data are true for just one line, as Example 5 (page 68) shows.

(c) If the given data are true for any two non-parallel lines, say l_1 and l_2, take one (say l_1) to be the x-axis. It can be shown (by methods of analytic geometry or linear algebra) that any line in the plane, for example the y-axis, can be expressed as a linear combination of two given non-parallel lines. Moreover, the projections y_n of P_n on the y-axis can be expressed in terms of the x_n and the projections z_n of P_n on the line l_2 [see Figure 4.15(b)], and the y_n have a limit y if the x_n and the z_n have limits. Thus the problem that P is the limit of the P_n can be reduced to the problem solved in (a).

CHAPTER FIVE

5.1 Assume that $\sqrt{2}$ is rational, i.e. that the diagonal of a unit square has length p_1/q_1 where p_1 and q_1 are integers. Then a square whose sides are q_1 units long has a diagonal of length p_1.

Now construct the following sequence of right isosceles triangles: The first has legs of length q_1 and a hypotenuse of length p_1, see Figure 5.4(b). Erect a perpendicular to the hypotenuse at a point which divides it into segments of lengths q_1 and $p_1 - q_1$. This perpendicular cuts off a corner of the first triangle, and this corner is our second triangle, clearly similar to the first, with leg of length q_2 and hypotenuse of length p_2. We observe [see Figure 5.4(b)] that

$$q_2 = p_1 - q_1 \quad \text{and} \quad p_2 = q_1 - q_2 = q_1 - (p_1 - q_1) = 2q_1 - p_1.$$

Now we repeat the construction and cut off the next corner triangle Its legs have length

$$q_3 = p_2 - q_2 = 2q_1 - p_1 - (p_1 - q_1) = 3q_1 - 2p_1$$

and its hypotenuse has length

$$p_3 = q_2 - q_3 = p_1 - q_1 - (3q_1 - 2p_1) = 3p_1 - 4q_1.$$

We continue cutting off corners, always obtaining an isosceles right triangle similar to all the previous ones. The leg of the nth triangle has length q_n, its hypotenuse has length p_n, and these lengths satify the relations

$$q_n = p_{n-1} - q_{n-1}, \qquad p_n = q_{n-1} - q_n.$$

Since $p_{n-1} = q_{n-2} - q_{n-1}$ we may express the length q_n of the nth leg by

$$q_n = q_{n-2} - 2q_{n-1}, \qquad\qquad n > 2,$$

that is, in terms of the lengths of the legs of the previous two triangles.

Now consider the sequence q_1, q_2, q_3, $\cdots$. Since p_1 and q_1 are integers, $q_2 = p_1 - q_1$ is an integer, $q_3 = q_1 - 2q_2$ is an integer and, in general, $q_n = q_{n-2} - 2q_{n-1}$ is an integer for all $n > 2$. It is clear from our construction that the legs of subsequent triangles decrease in length, i.e. that

$$q_1 > q_2 > q_3 > \cdots.$$

Thus the assumption that $\sqrt{2} = p_1/q_1$ is rational has led to an infinite decreasing sequence of positive integers, and no such sequence exists. We conclude that $\sqrt{2}$ is irrational.

In order to apply this method to $\sqrt{5}$, assume that $\sqrt{5} = r_1/s_1$ where r_1 and s_1 are integers. Blow up the rectangle of Figure 5.5 so that its sides are s_1, $2s_1$; then its diagonal is r_1. Our construction will lead to similar right triangles with legs s_n, $2s_n$ and hypotenuse r_n. The recursion relations will be

$$s_n = r_{n-1} - 2s_{n-1}, \qquad r_n = s_{n-1} - 2s_n,$$

so that

$$s_n = s_{n-2} - 2s_{n-1} - 2s_{n-1} = s_{n-2} - 4s_{n-1},$$

and the sequence s_1, s_2, s_3, $\cdots$ of lengths of shorter legs of the similar triangles is again a decreasing infinite sequence of integers.

These examples show how this method can be used to prove the irrationality of $\sqrt{k}$ for any integer k which can be written as the sum of the squares of two integers: $k = a^2 + b^2$. We have used it for $k = 1^2 + 1^2$, and for $k = 2^2 + 1^2$. The details of this generalization are left to the reader.

5.3   The kth fraction, F_k, is formed from the previous fraction, F_{k-1}, as follows:

$$F_k = \frac{1}{1 + F_{k-1}}.$$

If $\quad F_{k-1} = \dfrac{p}{q}$, $\quad$ then $\quad F_k = \dfrac{1}{1 + \dfrac{p}{q}} = \dfrac{q}{p + q}.$

5.4   (a) A sequence of finite parts of the expression $\sqrt{1 - \sqrt{1 - \sqrt{1 - \cdots}}}$ is formed in the following manner:

$$\sqrt{1}, \quad \sqrt{1 - \sqrt{1}}, \quad \sqrt{1 - \sqrt{1 - \sqrt{1}}}, \quad \cdots.$$

When we compute these numbers we see that this is the sequence 1, 0, 1, 0, $\cdots$, which has no limit.

(b) Since m satisfies $m^2 + m = 1$, its reciprocal satisfies

$$\frac{1}{\tau^2} + \frac{1}{\tau} = 1 \quad \text{or} \quad 1 + \tau = \tau^2.$$

The terms a_i of the sequence of finite parts

$$\sqrt{1}, \quad \sqrt{1 + \sqrt{1}}, \quad \sqrt{1 + \sqrt{1 + \sqrt{1}}}, \quad \cdots$$

obey the recursion formula

(1) $\qquad a_1 = \sqrt{1}, \qquad a_n = \sqrt{1 + a_{n-1}}, \quad \text{for } n = 2, 3, \cdots.$

We shall show that the increasing sequence a_1, a_2, $\cdots$ has a limit by applying the Bolzano-Weierstrass Theorem, see Section 3.8. In order to do this, we must find a bound B such that $a_1 < a_2 < \cdots < B$.

The fact that the a_i increase implies that $a_{n+1} - a_n > 0$ for $n = 1, 2, \cdots$. From (1) we have

$a_{n+1} - a_n$

$$= [\sqrt{1 + a_n} - \sqrt{1 + a_{n-1}}] \frac{\sqrt{1 + a_n} + \sqrt{1 + a_{n-1}}}{\sqrt{1 + a_n} + \sqrt{1 + a_{n-1}}}$$

$$= \frac{(1 + a_n) - (1 + a_{n-1})}{\sqrt{1 + a_n} + \sqrt{1 + a_{n-1}}} = \frac{a_n - a_{n-1}}{\sqrt{1 + a_n} + \sqrt{1 + a_{n-1}}}.$$

Since $a_i > 0$ for all i, the denominator in the last expression is greater than 2. Therefore

$$a_{n+1} - a_n < \frac{1}{2}(a_n - a_{n-1}) \qquad \text{for all } n,$$

and

(2) $\quad a_{n+1} - a_n < \dfrac{1}{2}(a_n - a_{n-1}) < \dfrac{1}{2}\left[\dfrac{1}{2}(a_{n-1} - a_{n-2})\right]$

$$< \cdots < \dfrac{1}{2^{n-1}}(a_2 - a_1).$$

Next we write a_{n+1} in the form

$$a_{n+1} = (a_{n+1} - a_n) + (a_n - a_{n-1}) + \cdots + (a_2 - a_1) + a_1$$

and apply inequality (2) to each expression in parentheses:

$$a_{n+1} < \dfrac{1}{2^{n-1}}(a_2 - a_1) + \dfrac{1}{2^{n-2}}(a_2 - a_1)$$

$$+ \cdots + (a_2 - a_1) + a_1$$

$$= (a_2 - a_1)\left[1 + \dfrac{1}{2} + \dfrac{1}{2^2} + \cdots + \dfrac{1}{2^{n-1}}\right] + a_1\,.$$

The expression in brackets never exceeds 2, so

$$a_{n+1} < 2(a_2 - a_1) + a_1 = 2(\sqrt{2} - 1) + 1 = 2\sqrt{2} - 1,$$

and this number bounds all terms of our sequence.

Observe that we did not make use of the fact that the limit of this sequence is $\tau = 1/m = 1 + m = 1.618 \cdots$. The bound

$$B = 2\sqrt{2} - 1 = 1.828 \cdots$$

which we constructed is somewhat larger than this limit.

5.5 The values of these ratios, calculated to six decimal places, are

$$\dfrac{5}{8} \approx .625000; \qquad \dfrac{8}{13} \approx .615385; \qquad \dfrac{13}{21} \approx .619048;$$

$$\dfrac{21}{34} \approx .617647; \qquad \dfrac{34}{55} \approx .618182; \qquad \dfrac{55}{89} \approx .617978.$$

The fractions

$$\dfrac{377}{610} \approx .618033 \qquad \text{and} \qquad \dfrac{610}{987} \approx .618034$$

are the 17th and 18th terms of the sequence.

5.6 Each successive fraction in the sequence is numerically closer to m. The approximate differences between m and the fractions $\frac{1}{2}$, $\frac{2}{3}$, $\frac{3}{5}$, $\frac{5}{8}$, $\frac{8}{13}$ are, respectively, .118034, .048633, .018034, .006966, and .002649. For a general proof of the fact that each convergent to an infinite continued fraction is closer to it than the previous convergent, see for example Chapter 3 (particularly Theorem 3.7) of the book by C. D. Olds, *Continued Fractions*, to appear in this series.

5.7 $m^5 = 2m - 3m^2 = 5m - 3;$ $m^6 = 5m^2 - 3m = 5 - 8m;$

$m^7 = 5m - 8m^2 = 13m - 8;$ $m^8 = 13m^2 - 8m = 13 - 21m;$ $\cdots$.

If f_n is the nth term of the Fibonacci Sequence, the formula for the nth power of m is

$$m^n = (-1)^n(f_{n-1} - f_n m).$$

The corresponding situation for τ is

$\tau^4 = 3\tau + 2;$ $\tau^5 = 5\tau + 3;$ $\tau^6 = 8\tau + 5;$ $\cdots$; $\tau^n = f_n\tau + f_{n-1}$.

5.8   A complete solution to Problem 5.8 is given in Chapter 6, pp. 114–117.

5.9   The way the vertices are ordered in successive rectangles reflects the fact that the shorter side of each rectangle (i.e. the line joining the 2 vertices named last in the ordering) is the longer side of the next one (i.e. the line between the vertices listed in the middle position). In each case the vertex named first is the one from which the 45° line is drawn to the point that is the first named vertex of the next rectangle.

5.10 The length of each successive segment of this zig-zag is the length of the preceding segment reduced by the factor $m < 1$. Hence, the length of the zig-zag is the sum of the infinite geometric series

$$\sqrt{2} + \sqrt{2} \cdot m + \sqrt{2} \cdot m^2 + \sqrt{2} \cdot m^3 + \cdots$$

$$= \frac{\sqrt{2}}{1 - m} = \frac{\sqrt{2}}{m^2} .$$

The solution to Problem 3.13 (p. 52) proves that the formula for the sum of *any* infinite geometric series with first term a and ratio $r < 1$ is $a/(1 - r)$.

5.11

Number of quarter-turns (degrees) about T		Distance from T to point on the spiral
1/6	(15°)	$AT \cdot m^{1/6}$
1/3	(30°)	$AT \cdot m^{1/3}$
1/2	(45°)	$AT \cdot m^{1/2}$
5/6	(75°)	$AT \cdot m^{5/6}$
4/3	(120°)	$AT \cdot m^{4/3}$
3/2	(135°)	$AT \cdot m^{3/2}$
5/3	(150°)	$AT \cdot m^{5/3}$
5/2	(225°)	$AT \cdot m^{5/2}$
$\cdots$	$\cdots$	$\cdots$
$\dfrac{2n + 1}{2}$	$\left(\dfrac{2n + 1}{2} \cdot 90°\right)$	$AT \cdot m^{(2n+1)/2}$

5.12 If t takes on negative values, we get a continuation of the spiral in a counter-clockwise direction from AT. As the values of t become smaller and smaller (i.e. as t approaches $-\infty$), R increases without limit.

5.13 To multiply a number R_1 by a number R_2 by means of the spiral in Figure 5.11 (where the distance AT is now taken to be the unit of measurement) we use the ruler to locate those points P_1 and P_2 on the spiral which have distances R_1 and R_2 from T:

$$P_1T = R_1, \qquad P_2T = R_2.$$

Now we follow the spiral from the point A to the point P_1 and denote by α_1 the angle through which the radius vector to the spiral must rotate to get from TA to TP_1. (Observe that if $R_1 < 1$, then we reach P_1 by going in the clockwise direction and α_1 will be taken to be positive; if $R_1 > 1$, then we reach P_1 by going counter-clockwise and α_1 will be taken negative.) Next, we follow the spiral from A to P_2 and measure the angle α_2 by which the radius vector must be rotated to get from TA to TP_2. Now we add the angles α_1 and α_2, and rotate the line AT through the angle $\alpha_1 + \alpha_2$ always following the spiral from the point A on. This will lead to a point P_3 on the spiral whose distance from T is

$$P_3T = R_1 \cdot R_2.$$

This method is just a geometric interpretation of the law of exponents: Given

$$R_1 = m^{\alpha_1}, \qquad R_2 = m^{\alpha_2},$$

we have found

$$R_1 \cdot R_2 = m^{\alpha_1 + \alpha_2}.$$

5.14 If the radius vectors TP_1, TP_2, $\cdots$, TP_n have the same direction but different magnitudes R_1, R_2, $\cdots$, R_n, then the angles of rotation α_1, α_2, $\cdots$, α_n, measured from the line through T and A as this line passes through each point of the curve from A to P_1, to P_2, $\cdots$, to P_n, differ only by multiples of 2π radians (one full turn about T, i.e. 4 quarter-turns). This property corresponds to the fact that the logarithms of the numbers represented by the lengths R_1, R_2, $\cdots$, R_n would differ only in their characteristics, i.e. in the integer part of the logarithm. (If α is measured in quarter-turns, these logarithms would differ by multiples of 4.) If α is between $4k$ and $4(k + 1)$ quarter-turns, $\alpha - 4k$ would correspond to the mantissa and would determine the direction of the line TP, while the characteristic $4k$ would determine on which "ring" of the spiral the point P lies.

CHAPTER SIX

6.1   Assume to the contrary that there exist integers a and b such that

$$\frac{a}{b} \cdot (1 + \sqrt{2}) = 1.$$

Then

$$1 + \sqrt{2} = \frac{b}{a},$$

or

$$\sqrt{2} = \frac{b}{a} - 1 = \frac{b - a}{a}.$$

But if b and a are integers, $b - a$ is also an integer, and the last equality states that $\sqrt{2}$ is rational, which is false. Therefore the reciprocal of $1 + \sqrt{2}$ is not rational.

6.2   Let d be the highest common factor of a and b, and let x and y be integers. Then the integer

$$ax + by = c = a'dx + b'dy = d(a'x + b'y)$$

is clearly divisible by d.

Conversely, if c is divisible by d, the highest common factor of a and b, then we can find integers x and y such that

$$ax + by = c$$

in the following way. We divide the equation by d obtaining

$$a'x + b'y = c',$$

where a' and b' are relatively prime. In this case it is known (see e.g. the discussion of Euclid's algorithm in *Continued Fractions* by C. D. Olds, to appear in this series) that there exist integers x_1 and y_1 such that

$$a'x_1 + b'y_1 = 1;$$

then the integers $x = c'x_1$, $y = c'y_1$ will satisfy

$$a'x + b'y = c',$$

and hence also $ax + by = c$.

6.3 2^{n-1}, $n = 1, 2, 3, \cdots$.

6.4   When $N = 1$, N and the sum of its digits clearly have the same residue modulo 3. This proves the first step in the induction.

Suppose that k is an integer such that k and the sum of its digits have the same residue modulo 3, i.e. such that

$$k = 3q + r, \qquad\qquad 0 \le r < 3,$$

and the sum of the digits of k is given by

$$3s + r, \qquad\qquad 0 \le r < 3.$$

To prove the inductive step, we must show that $k + 1$ and the sum of its digits have the same remainder when divided by 3. When $0 \le r + 1 < 3$, we have

$$k + 1 = 3q + (r + 1);$$

otherwise

$$k + 1 = 3(q + 1).$$

If the last m $(m \ge 0)$ digits of a number k are all 9's, these 9's will become 0's when 1 is added to k, but the first digit which is not a 9 will be increased by 1. Since the sum of the digits of k is $3s + r$, we may write the sum of the digits of $k + 1$ in the form

$$(3s + r) + 1 - 9m,$$

which is equivalent to

$$3(s - 3m) + (r + 1).$$

Thus $k + 1$ and the sum of the digits of $k + 1$ have the same residue modulo 3.

6.5 The assertion is true for $n = 1$. Assume that for $n = k$,

$$1 + 2 + \cdots + k = \tfrac{1}{2}k(k + 1),$$

and consider the case for $n = k + 1$. By applying the inductive hypothesis we get

$$1 + 2 + \cdots + k + k + 1 = \tfrac{1}{2}k(k + 1) + k + 1,$$

which can be written

$$\tfrac{1}{2}k^2 + \tfrac{3}{2}k + 1 = \tfrac{1}{2}(k^2 + 3k + 2) = \tfrac{1}{2}(k + 1)(k + 2).$$

Since this is of the form $\tfrac{1}{2}n(n + 1)$, the proof is complete.

6.6 For all integers n, 2^{n+9} exceeds $(n + 9)^3$.

6.7 It is true that, when k is an integer greater than 2, then

$$2k^2 > k^2 + 2k + 2 > (k + 1)^2;$$

to show this, note that when $k > 2$, then $k - 2 > 0$, and since k is an integer, $k - 2 \ge 1$, $k \ge 3$ so that $k(k - 2) > 1$ or $k^2 > 2k + 1$. Hence

$$2k^2 = k^2 + k^2 > k^2 + 2k + 1,$$

that is, $2k^2 > (k + 1)^2$.

This fact does not enable us to prove that 2^n exceeds n^2 for all $n > 2$ because, in order to use that $2^{k+1} > 2k^2$, we had to assume that $2^k > k^2$, and it is *not* true that 2^3 exceeds 3^2.

6.8 (a) For $N = 1$ we have $(2^1)^1 = 2 = 2^{(1^2)}$. If, when $N = k$,

$$(2^k)^k = 2^{(k^2)},$$

then

$$(2^{k+1})^{k+1} = (2^k \cdot 2)^{k+1} = (2^k \cdot 2)^k (2^k \cdot 2)$$

$$= (2^k)^k \cdot 2^{2k} \cdot 2$$

$$= 2^{(k^2 + 2k + 1)}$$

$$= 2^{[(k+1)^2]}$$

(b) From $2^N > N^2$ if $N > 4$ we get

$$(2^N)^N > (N^2)^N \quad \text{or} \quad 2^{(N^2)} > (N^2)^N.$$

Thus, if we take $n = N^2$, we have that for $N > 4$, $2^n > n^N$. But from the proofs of Theorems 2 and 3 we know that for $N = 2$ and $N = 3$, $2^n > n^N$ only if n exceeds N^2. This suggests that we can prove the inductive step of Therorem N by showing that for all $n > N^2$, $2^n > n^N$.

6.9 If $2^k > k^N$, and $k > N^2$, then

$$2^{k+1} = 2 \cdot 2^k > 2k^N = k^N + k^N > k^N + N^2 k^{N-1}$$

so that

$$2^{k+1} > k^N + Nk^{N-1} + N(N-1)k^{N-1}$$

$$\geq k^N + Nk^{N-1} + N(N-1)k^{N-2}$$

$$= k^N + Nk^{N-1} + \frac{N(N-1)}{2} k^{N-2} + \frac{N(N-1)}{2} k^{N-2}$$

$$\geq k^N + Nk^{N-1} + \frac{N(N-1)}{2} k^{N-2} + \frac{N(N-1)(N-2)}{3} k^{N-3}$$

$$= k^N + Nk^{N-1} + \frac{N(N-1)}{2} k^{N-2} + \frac{N(N-1)(N-2)}{2 \cdot 3} k^{N-3}$$

$$+ \frac{N(N-1)(N-2)}{2 \cdot 3} k^{N-3}$$

$$\geq \cdots$$

$$\geq k^N + Nk^{N-1} + \frac{N(N-1)}{2!} k^{N-2} + \frac{N(N-1)(N-2)}{3!} k^{N-3}$$

$$+ \cdots + \frac{N(N-1) \cdots [N - (N-2)]}{(N-1)!} k^{N-(N-1)} + 1$$

$$= (k+1)^N.$$

6.10 We have proved (Theorem 1) that when $N = 1$,

$$2^n > n^N = n \qquad \text{for all integers } n.$$

Assume that when $N = k$ and n exceeds k^2, it is true that $2^n > n^k$. It follows from Problem 6.9 that $2^n > n^{k+1}$ provided that $n > (k+1)^2$, which is all we need to complete the proof that $2^n > n^N$ for all integers N and n such that $n > N^2$.

6 11 Let us try to imitate the proof of Lagrange's Theorem (pages 115–117) in the present case and let us observe what modifications will be necessary.

The box principle tells us that any sequence of residues (mod N) has a repeating consecutive pair within $N^2 + 2$ terms. If the pairs a_i, a_{i+1} and a_k, a_{k+1} have the same residues, then from

$$a_i \equiv a_k (\text{mod } N) \qquad \text{and} \qquad a_{i+1} \equiv a_{k+1} (\text{mod } N)$$

we get

$$3a_i \equiv 3a_k (\text{mod } N) \qquad \text{and} \qquad 2a_{i+1} \equiv 2a_{k+1} (\text{mod } N).$$

It follows that

$$2a_{i+1} + 3a_i \equiv 2a_{k+1} + 3a_k (\text{mod } N),$$

or

$$a_{i+2} \equiv a_{k+2} (\text{mod } N).$$

By the same argument

$$a_{i+3} \equiv a_{k+3} (\text{mod } N),$$
$$a_{i+4} \equiv a_{k+4} (\text{mod } N),$$
$$\dots\dots\dots\dots\dots\dots ,$$

which shows that the sequence of residues (mod N) of the sequence given by $a_{n+1} = 2a_n + 3a_{n-1}$ is periodic.

Let the period of the sequence be p. Then

$$a_j \equiv a_{j+p} (\text{mod } N) \qquad\qquad (j \geq T)$$

from some j on, say $j = T$. Suppose a_T is not the first term of the sequence. From the recursion formula we have

$$3a_{T-1} = a_{T+1} - 2a_T$$
$$\equiv a_{T+1+p} - 2a_{T+p} (\text{mod } N)$$
$$= 3a_{T+p-1} .$$

Hence,

$$3a_{T-1} \equiv 3a_{T-1+p} (\text{mod } N),$$

or

$$3(a_{T-1} - a_{T-1+p}) \equiv 0 (\text{mod } N).$$

Clearly, we can conclude that

$$a_{T-1} \equiv a_{T-1+p} \pmod{N}$$

only if we assume that 3 and N are relatively prime; otherwise, it does not necessarily follow that N is a factor of $a_{T-1} - a_{T-1+p}$. Since T is a a finite integer, this process applied successively to $T - 1$, $T - 2$, $T - 3$, $\cdots$, eventually must lead to

$$a_1 \equiv a_{1+p} \pmod{N}.$$

Thus, the sequence of residues (mod N), $N \geq 2$, of the sequence defined by $a_{n+1} = 2a_n + 3a_{n-1}$, $n \geq 2$ (where the initial values a_1 and a_2 may be any given integers) is periodic. If 3 and N are relatively prime, then the periodic part begins with the residue of a_1.

6.12 The residues (mod$N$), $N \geq 2$, of any sequence defined by

$$a_{n+1} = \alpha a_n + \beta a_{n-1}$$

with arbitrary initial integers a_1 and a_2 are periodic; the repetition of a pair occurs within at most $N^2 + 2$ terms. If N and β are relatively prime, then the periodic part begins with the residue of a_1.

6.13 The sequence of residues (mod$N$) of any sequence defined by

$$a_{n+1} = \alpha a_n + \beta a_{n-1} + \gamma a_{n-2}, \qquad\qquad n \geq 3,$$

has a repeated consecutive triplet within $N^3 + 3$ terms. If N and γ are relatively prime, the sequence of residues (mod$N$) is periodic from the beginning on.

6.14 In general, the sequence of residues (mod$N$) of a sequence a_1, a_2, $\cdots$, a_n, $\cdots$ built (after the nth term) on a rule expressing the $(n + 1)$th term as a linear combination of the preceding n terms is periodic from the beginning on and will repeat within $N^n + n$ terms whenever N has no factors greater than 1 in common with the coefficient of the earliest term in the recursion formula.

6.15 The lines $x = a$ for all rational a constitute a countable infinity of lines since the set of all rational numbers is countable. The same is true for the sets $y = b$, $x = c$, $y = d$ for rational b, c, d. Since all special rectangles are formed by combining 4 sides, each from one of these sets, we obtain $\aleph_0 \cdot \aleph_0 \cdot \aleph_0 \cdot \aleph_0 = \aleph_0$ possible special rectangles. (This even includes the degenerate rectangles in which a pair of opposite sides coincides. Therefore, the non-degenerate special rectangles certainly constitute a countable set.)

6.16 Let P be the point (x_0, y_0), and let d be the minimum distance from P to any point on the given rectangle. Then there exist rational numbers $\delta_1 < \frac{1}{4}d$ and $\delta_2 < \frac{1}{4}d$ such that

$$x = x_0 + \delta_1 = a \qquad\text{and}\qquad x = x_0 - \delta_2 = a'$$

are rational, and numbers $\epsilon_1 < \frac{1}{4}d$ and $\epsilon_2 < \frac{1}{4}d$ such that

$$y = y_0 + \epsilon_1 = b \qquad\text{and}\qquad y = y_0 - \epsilon_2 = b'$$

are rational. It follows that the sides of a special rectangle R lie on the lines $x = a$, $x = a'$, $y = b$, $y = b'$, and the point P is inside of this rectangle. Furthermore, since the length of the diagonal of R is

$$\sqrt{(\delta_1 + \delta_2)^2 + (\epsilon_1 + \epsilon_2)^2} < d,$$

the distance from P to the farthest point on R is less than the minimum distance from P to the given rectangle. Hence, the special rectangle lies entirely within the given one.

6.17 If we assume that there is no point P in the set X such that every rectangle containing P contains uncountably many points of X, then every point in the set X must be inside at least one rectangle containing a countable set of points of X. In this case, the solution to Problem 6.16 shows that every point of X is inside of a special rectangle which is entirely within the rectangle containing a countable set of points belonging to X, and so also contains at most a countable infinity of points of X. Now, we have proved (Problem 6.15) that the set of all special rectangles is of power $\aleph_0$; therefore, the set of special rectangles with which we are concerned is certainly countable. Moreover, since each of these special rectangles contains a countable set of points of X, it follows from $\aleph_0 \cdot \aleph_0 = \aleph_0$ that the set X is countable. But this contradicts the hypothesis that the given set is uncountable; hence, there must exist some point P in X such that every rectangle containing P contains uncountably many points belonging to the set X.

6.18 Take the point P obtained in the solution to Problem 6.17, and a sequence of decreasing intervals (rectangles in the case of the plane) closing down on P. In each of these intervals pick one point of X from among the uncountably many which are available. This gives a sequence P_1, P_2, P_3, $\cdots$ of points of X which form the desired convergent sequence. A proof such as this is called "non-constructive" because no mechanism is provided for actually defining each point P_n. Since we know nothing about X except that it is uncountable, no method of selection is available to us.

Bibliography

Cajori, Florian, *A History of Mathematics*, New York and London: The Macmillan Company, 1919.

Courant, Richard, *Differential and Integral Calculus*, Vol. 1, New York: Nordemann, 1940.

Courant, Richard and Robbins, Herbert, *What is Mathematics?*, London and New York: Oxford University Press, 1941.

Coxeter, H. S. M., *Introduction to Geometry*, New York: John Wiley and Sons, Inc., 1961.

Davis, Philip J., *The Lore of Large Numbers*, Vol. 6, *New Mathematical Library*, New York and New Haven: Random House, Inc. and Yale University, 1961.

Eves, Howard, *An Introduction to the History of Mathematics*, New York: Rhinehart and Company, 1958.

Frankel, Abraham A., *Abstract Set Theory*, Amsterdam, Holland: North-Holland Publishing Company, 1953.

Hardy, Godfrey H., *A Course of Pure Mathematics*, Cambridge: Cambridge University Press, 1938.

Heath, Thomas L., *Manual of Greek Mathematics*, London and New York: Oxford University Press, 1931.

Hilbert, David and Cohn-Vossen, Stephan, *Geometry and the Imagination*, New York: Chelsea, 1952.

Neugebauer, Otto, *The Exact Sciences in Antiquity*, Copenhagen, Denmark and Princeton, New Jersey: E. Munksgaard and Princeton University Press, 1951.

Niven, Ivan, *Numbers: Rational and Irrational*, Vol. 1, *New Mathematical Library*, New York and New Haven: Random House, Inc. and Yale University, 1961.

Olds, C. D., *Continued Fractions*, to be published for the *New Mathematical Library*, approximately 1962.

Rademacher, Hans and Toeplitz, Otto, *The Enjoyment of Mathematics*, Princeton: Princeton University Press, 1957.

Steinhaus, Hugo, *Mathematical Snapshots*, New York: G. E. Stechert and Company, 1938 (2nd ed., London and New York: Oxford University Press, 1960).

Wilder, Raymond L., *Introduction to the Foundations of Mathematics*, New York: John Wiley and Sons, Inc., 1952.

Yaglom, I. M., *Geometric Transformations*, translated from the Russian by Allen Shields, to be published for the *New Mathematical Library*, approximately 1962.

A CATALOG OF SELECTED
DOVER BOOKS
IN ALL FIELDS OF INTEREST

A CATALOG OF SELECTED DOVER
BOOKS IN ALL FIELDS OF INTEREST

CONCERNING THE SPIRITUAL IN ART, Wassily Kandinsky. Pioneering work by father of abstract art. Thoughts on color theory, nature of art. Analysis of earlier masters. 12 illustrations. 80pp. of text. 5⅜ x 8½. 23411-8 Pa. $4.95

ANIMALS: 1,419 Copyright-Free Illustrations of Mammals, Birds, Fish, Insects, etc., Jim Harter (ed.). Clear wood engravings present, in extremely lifelike poses, over 1,000 species of animals. One of the most extensive pictorial sourcebooks of its kind. Captions. Index. 284pp. 9 x 12. 23766-4 Pa. $14.95

CELTIC ART: The Methods of Construction, George Bain. Simple geometric techniques for making Celtic interlacements, spirals, Kells-type initials, animals, humans, etc. Over 500 illustrations. 160pp. 9 x 12. (USO) 22923-8 Pa. $9.95

AN ATLAS OF ANATOMY FOR ARTISTS, Fritz Schider. Most thorough reference work on art anatomy in the world. Hundreds of illustrations, including selections from works by Vesalius, Leonardo, Goya, Ingres, Michelangelo, others. 593 illustrations. 192pp. 7⅛ x 10¼. 20241-0 Pa. $9.95

CELTIC HAND STROKE-BY-STROKE (Irish Half-Uncial from "The Book of Kells"): An Arthur Baker Calligraphy Manual, Arthur Baker. Complete guide to creating each letter of the alphabet in distinctive Celtic manner. Covers hand position, strokes, pens, inks, paper, more. Illustrated. 48pp. 8¼ x 11. 24336-2 Pa. $3.95

EASY ORIGAMI, John Montroll. Charming collection of 32 projects (hat, cup, pelican, piano, swan, many more) specially designed for the novice origami hobbyist. Clearly illustrated easy-to-follow instructions insure that even beginning papercrafters will achieve successful results. 48pp. 8¼ x 11. 27298-2 Pa. $3.50

THE COMPLETE BOOK OF BIRDHOUSE CONSTRUCTION FOR WOODWORKERS, Scott D. Campbell. Detailed instructions, illustrations, tables. Also data on bird habitat and instinct patterns. Bibliography. 3 tables. 63 illustrations in 15 figures. 48pp. 5¼ x 8½. 24407-5 Pa. $2.50

BLOOMINGDALE'S ILLUSTRATED 1886 CATALOG: Fashions, Dry Goods and Housewares, Bloomingdale Brothers. Famed merchants' extremely rare catalog depicting about 1,700 products: clothing, housewares, firearms, dry goods, jewelry, more. Invaluable for dating, identifying vintage items. Also, copyright-free graphics for artists, designers. Co-published with Henry Ford Museum & Greenfield Village. 160pp. 8¼ x 11. 25780-0 Pa. $10.95

HISTORIC COSTUME IN PICTURES, Braun & Schneider. Over 1,450 costumed figures in clearly detailed engravings–from dawn of civilization to end of 19th century. Captions. Many folk costumes. 256pp. 8⅜ x 11¾. 23150-X Pa. $12.95

STICKLEY CRAFTSMAN FURNITURE CATALOGS, Gustav Stickley and L. & J. G. Stickley. Beautiful, functional furniture in two authentic catalogs from 1910. 594 illustrations, including 277 photos, show settles, rockers, armchairs, reclining chairs, bookcases, desks, tables. 183pp. 6½ x 9¼. 23838-5 Pa. $11.95

AMERICAN LOCOMOTIVES IN HISTORIC PHOTOGRAPHS: 1858 to 1949, Ron Ziel (ed.). A rare collection of 126 meticulously detailed official photographs, called "builder portraits," of American locomotives that majestically chronicle the rise of steam locomotive power in America. Introduction. Detailed captions. xi + 129pp. 9 x 12. 27393-8 Pa. $13.95

AMERICA'S LIGHTHOUSES: An Illustrated History, Francis Ross Holland, Jr. Delightfully written, profusely illustrated fact-filled survey of over 200 American lighthouses since 1716. History, anecdotes, technological advances, more. 240pp. 8 x 10¾. 25576-X Pa. $12.95

TOWARDS A NEW ARCHITECTURE, Le Corbusier. Pioneering manifesto by founder of "International School." Technical and aesthetic theories, views of industry, economics, relation of form to function, "mass-production split" and much more. Profusely illustrated. 320pp. 6⅛ x 9¼. (USO) 25023-7 Pa. $9.95

HOW THE OTHER HALF LIVES, Jacob Riis. Famous journalistic record, exposing poverty and degradation of New York slums around 1900, by major social reformer. 100 striking and influential photographs. 233pp. 10 x 7⅞. 22012-5 Pa. $11.95

FRUIT KEY AND TWIG KEY TO TREES AND SHRUBS, William M. Harlow. One of the handiest and most widely used identification aids. Fruit key covers 120 deciduous and evergreen species; twig key 160 deciduous species. Easily used. Over 300 photographs. 126pp. 5⅜ x 8½. 20511-8 Pa. $3.95

COMMON BIRD SONGS, Dr. Donald J. Borror. Songs of 60 most common U.S. birds: robins, sparrows, cardinals, bluejays, finches, more—arranged in order of increasing complexity. Up to 9 variations of songs of each species. Cassette and manual 99911-4 $8.95

ORCHIDS AS HOUSE PLANTS, Rebecca Tyson Northen. Grow cattleyas and many other kinds of orchids—in a window, in a case, or under artificial light. 63 illustrations. 148pp. 5⅜ x 8½. 23261-1 Pa. $5.95

MONSTER MAZES, Dave Phillips. Masterful mazes at four levels of difficulty. Avoid deadly perils and evil creatures to find magical treasures. Solutions for all 32 exciting illustrated puzzles. 48pp. 8¼ x 11. 26005-4 Pa. $2.95

MOZART'S DON GIOVANNI (DOVER OPERA LIBRETTO SERIES), Wolfgang Amadeus Mozart. Introduced and translated by Ellen H. Bleiler. Standard Italian libretto, with complete English translation. Convenient and thoroughly portable—an ideal companion for reading along with a recording or the performance itself. Introduction. List of characters. Plot summary. 121pp. 5¼ x 8½. 24944-1 Pa. $3.95

TECHNICAL MANUAL AND DICTIONARY OF CLASSICAL BALLET, Gail Grant. Defines, explains, comments on steps, movements, poses and concepts. 15-page pictorial section. Basic book for student, viewer. 127pp. 5⅜ x 8½. 21843-0 Pa. $4.95

THE CLARINET AND CLARINET PLAYING, David Pino. Lively, comprehensive work features suggestions about technique, musicianship, and musical interpretation, as well as guidelines for teaching, making your own reeds, and preparing for public performance. Includes an intriguing look at clarinet history. "A godsend," The Clarinet, Journal of the International Clarinet Society. Appendixes. 7 illus. 320pp. 5⅜ x 8½. 40270-3 Pa. $9.95

HOLLYWOOD GLAMOR PORTRAITS, John Kobal (ed.). 145 photos from 1926-49. Harlow, Gable, Bogart, Bacall; 94 stars in all. Full background on photographers, technical aspects. 160pp. 8⅜ x 11¼. 23352-9 Pa. $12.95

THE ANNOTATED CASEY AT THE BAT: A Collection of Ballads about the Mighty Casey/Third, Revised Edition, Martin Gardner (ed.). Amusing sequels and parodies of one of America's best-loved poems: Casey's Revenge, Why Casey Whiffed, Casey's Sister at the Bat, others. 256pp. 5⅜ x 8½. 28598-7 Pa. $8.95

THE RAVEN AND OTHER FAVORITE POEMS, Edgar Allan Poe. Over 40 of the author's most memorable poems: "The Bells," "Ulalume," "Israfel," "To Helen," "The Conqueror Worm," "Eldorado," "Annabel Lee," many more. Alphabetic lists of titles and first lines. 64pp. 5$\frac{3}{16}$ x 8¼. 26685-0 Pa. $1.00

PERSONAL MEMOIRS OF U. S. GRANT, Ulysses Simpson Grant. Intelligent, deeply moving firsthand account of Civil War campaigns, considered by many the finest military memoirs ever written. Includes letters, historic photographs, maps and more. 528pp. 6⅛ x 9¼. 28587-1 Pa. $12.95

ANCIENT EGYPTIAN MATERIALS AND INDUSTRIES, A. Lucas and J. Harris. Fascinating, comprehensive, thoroughly documented text describes this ancient civilization's vast resources and the processes that incorporated them in daily life, including the use of animal products, building materials, cosmetics, perfumes and incense, fibers, glazed ware, glass and its manufacture, materials used in the mummification process, and much more. 544pp. 6⅛ x 9¼. (USO) 40446-3 Pa. $16.95

RUSSIAN STORIES/PYCCKNE PACCKA3bl: A Dual-Language Book, edited by Gleb Struve. Twelve tales by such masters as Chekhov, Tolstoy, Dostoevsky, Pushkin, others. Excellent word-for-word English translations on facing pages, plus teaching and study aids, Russian/English vocabulary, biographical/critical introductions, more. 416pp. 5⅜ x 8½. 26244-8 Pa. $9.95

PHILADELPHIA THEN AND NOW: 60 Sites Photographed in the Past and Present, Kenneth Finkel and Susan Oyama. Rare photographs of City Hall, Logan Square, Independence Hall, Betsy Ross House, other landmarks juxtaposed with contemporary views. Captures changing face of historic city. Introduction. Captions. 128pp. 8¼ x 11. 25790-8 Pa. $9.95

AIA ARCHITECTURAL GUIDE TO NASSAU AND SUFFOLK COUNTIES, LONG ISLAND, The American Institute of Architects, Long Island Chapter, and the Society for the Preservation of Long Island Antiquities. Comprehensive, well-researched and generously illustrated volume brings to life over three centuries of Long Island's great architectural heritage. More than 240 photographs with authoritative, extensively detailed captions. 176pp. 8¼ x 11. 26946-9 Pa. $14.95

NORTH AMERICAN INDIAN LIFE: Customs and Traditions of 23 Tribes, Elsie Clews Parsons (ed.). 27 fictionalized essays by noted anthropologists examine religion, customs, government, additional facets of life among the Winnebago, Crow, Zuni, Eskimo, other tribes. 480pp. 6⅛ x 9¼. 27377-6 Pa. $10.95

FRANK LLOYD WRIGHT'S DANA HOUSE, Donald Hoffmann. Pictorial essay of residential masterpiece with over 160 interior and exterior photos, plans, elevations, sketches and studies. 128pp. 9¼ x 10¾. 29120-0 Pa. $12.95

THE MALE AND FEMALE FIGURE IN MOTION: 60 Classic Photographic Sequences, Eadweard Muybridge. 60 true-action photographs of men and women walking, running, climbing, bending, turning, etc., reproduced from rare 19th-century masterpiece. vi + 121pp. 9 x 12. 24745-7 Pa. $10.95

1001 QUESTIONS ANSWERED ABOUT THE SEASHORE, N. J. Berrill and Jacquelyn Berrill. Queries answered about dolphins, sea snails, sponges, starfish, fishes, shore birds, many others. Covers appearance, breeding, growth, feeding, much more. 305pp. 5¼ x 8¼. 23366-9 Pa. $9.95

ATTRACTING BIRDS TO YOUR YARD, William J. Weber. Easy-to-follow guide offers advice on how to attract the greatest diversity of birds: birdhouses, feeders, water and waterers, much more. 96pp. 5³⁄₁₆ x 8¼. 28927-3 Pa. $2.50

MEDICINAL AND OTHER USES OF NORTH AMERICAN PLANTS: A Historical Survey with Special Reference to the Eastern Indian Tribes, Charlotte Erichsen-Brown. Chronological historical citations document 500 years of usage of plants, trees, shrubs native to eastern Canada, northeastern U.S. Also complete identifying information. 343 illustrations. 544pp. 6½ x 9¼. 25951-X Pa. $12.95

STORYBOOK MAZES, Dave Phillips. 23 stories and mazes on two-page spreads: Wizard of Oz, Treasure Island, Robin Hood, etc. Solutions. 64pp. 8¼ x 11. 23628-5 Pa. $2.95

AMERICAN NEGRO SONGS: 230 Folk Songs and Spirituals, Religious and Secular, John W. Work. This authoritative study traces the African influences of songs sung and played by black Americans at work, in church, and as entertainment. The author discusses the lyric significance of such songs as "Swing Low, Sweet Chariot," "John Henry," and others and offers the words and music for 230 songs. Bibliography. Index of Song Titles. 272pp. 6½ x 9¼. 40271-1 Pa. $9.95

MOVIE-STAR PORTRAITS OF THE FORTIES, John Kobal (ed.). 163 glamor, studio photos of 106 stars of the 1940s: Rita Hayworth, Ava Gardner, Marlon Brando, Clark Gable, many more. 176pp. 8⅜ x 11¼. 23546-7 Pa. $14.95

BENCHLEY LOST AND FOUND, Robert Benchley. Finest humor from early 30s, about pet peeves, child psychologists, post office and others. Mostly unavailable elsewhere. 73 illustrations by Peter Arno and others. 183pp. 5⅜ x 8½. 22410-4 Pa. $6.95

YEKL and THE IMPORTED BRIDEGROOM AND OTHER STORIES OF YIDDISH NEW YORK, Abraham Cahan. Film Hester Street based on Yekl (1896). Novel, other stories among first about Jewish immigrants on N.Y.'s East Side. 240pp. 5⅜ x 8½. 22427-9 Pa. $6.95

SELECTED POEMS, Walt Whitman. Generous sampling from *Leaves of Grass.* Twenty-four poems include "I Hear America Singing," "Song of the Open Road," "I Sing the Body Electric," "When Lilacs Last in the Dooryard Bloom'd," "O Captain! My Captain!"–all reprinted from an authoritative edition. Lists of titles and first lines. 128pp. 5³⁄₁₆ x 8¼. 26878-0 Pa. $1.00

THE BEST TALES OF HOFFMANN, E. T. A. Hoffmann. 10 of Hoffmann's most important stories: "Nutcracker and the King of Mice," "The Golden Flowerpot," etc. 458pp. 5⅜ x 8½. 21793-0 Pa. $9.95

FROM FETISH TO GOD IN ANCIENT EGYPT, E. A. Wallis Budge. Rich detailed survey of Egyptian conception of "God" and gods, magic, cult of animals, Osiris, more. Also, superb English translations of hymns and legends. 240 illustrations. 545pp. 5⅜ x 8½. 25803-3 Pa. $13.95

FRENCH STORIES/CONTES FRANÇAIS: A Dual-Language Book, Wallace Fowlie. Ten stories by French masters, Voltaire to Camus: "Micromegas" by Voltaire; "The Atheist's Mass" by Balzac; "Minuet" by de Maupassant; "The Guest" by Camus, six more. Excellent English translations on facing pages. Also French-English vocabulary list, exercises, more. 352pp. 5⅜ x 8½. 26443-2 Pa. $9.95

CHICAGO AT THE TURN OF THE CENTURY IN PHOTOGRAPHS: 122 Historic Views from the Collections of the Chicago Historical Society, Larry A. Viskochil. Rare large-format prints offer detailed views of City Hall, State Street, the Loop, Hull House, Union Station, many other landmarks, circa 1904-1913. Introduction. Captions. Maps. 144pp. 9⅜ x 12¼. 24656-6 Pa. $12.95

OLD BROOKLYN IN EARLY PHOTOGRAPHS, 1865-1929, William Lee Younger. Luna Park, Gravesend race track, construction of Grand Army Plaza, moving of Hotel Brighton, etc. 157 previously unpublished photographs. 165pp. 8⅞ x 11¾. 23587-4 Pa. $13.95

THE MYTHS OF THE NORTH AMERICAN INDIANS, Lewis Spence. Rich anthology of the myths and legends of the Algonquins, Iroquois, Pawnees and Sioux, prefaced by an extensive historical and ethnological commentary. 36 illustrations. 480pp. 5⅜ x 8½. 25967-6 Pa. $10.95

AN ENCYCLOPEDIA OF BATTLES: Accounts of Over 1,560 Battles from 1479 B.C. to the Present, David Eggenberger. Essential details of every major battle in recorded history from the first battle of Megiddo in 1479 B.C. to Grenada in 1984. List of Battle Maps. New Appendix covering the years 1967-1984. Index. 99 illustrations. 544pp. 6½ x 9¼. 24913-1 Pa. $16.95

SAILING ALONE AROUND THE WORLD, Captain Joshua Slocum. First man to sail around the world, alone, in small boat. One of great feats of seamanship told in delightful manner. 67 illustrations. 294pp. 5⅜ x 8½. 20326-3 Pa. $6.95

ANARCHISM AND OTHER ESSAYS, Emma Goldman. Powerful, penetrating, prophetic essays on direct action, role of minorities, prison reform, puritan hypocrisy, violence, etc. 271pp. 5⅜ x 8½. 22484-8 Pa. $7.95

MYTHS OF THE HINDUS AND BUDDHISTS, Ananda K. Coomaraswamy and Sister Nivedita. Great stories of the epics; deeds of Krishna, Shiva, taken from puranas, Vedas, folk tales; etc. 32 illustrations. 400pp. 5⅜ x 8½. 21759-0 Pa. $12.95

THE TRAUMA OF BIRTH, Otto Rank. Rank's controversial thesis that anxiety neurosis is caused by profound psychological trauma which occurs at birth. 256pp. 5⅜ x 8½. 27974-X Pa. $7.95

A THEOLOGICO-POLITICAL TREATISE, Benedict Spinoza. Also contains unfinished Political Treatise. Great classic on religious liberty, theory of government on common consent. R. Elwes translation. Total of 421pp. 5⅜ x 8½. 20249-6 Pa. $9.95

MY BONDAGE AND MY FREEDOM, Frederick Douglass. Born a slave, Douglass became outspoken force in antislavery movement. The best of Douglass' autobiographies. Graphic description of slave life. 464pp. 5⅜ x 8½. 22457-0 Pa. $8.95

FOLLOWING THE EQUATOR: A Journey Around the World, Mark Twain. Fascinating humorous account of 1897 voyage to Hawaii, Australia, India, New Zealand, etc. Ironic, bemused reports on peoples, customs, climate, flora and fauna, politics, much more. 197 illustrations. 720pp. 5⅜ x 8½. 26113-1 Pa. $15.95

THE PEOPLE CALLED SHAKERS, Edward D. Andrews. Definitive study of Shakers: origins, beliefs, practices, dances, social organization, furniture and crafts, etc. 33 illustrations. 351pp. 5⅜ x 8½. 21081-2 Pa. $8.95

THE MYTHS OF GREECE AND ROME, H. A. Guerber. A classic of mythology, generously illustrated, long prized for its simple, graphic, accurate retelling of the principal myths of Greece and Rome, and for its commentary on their origins and significance. With 64 illustrations by Michelangelo, Raphael, Titian, Rubens, Canova, Bernini and others. 480pp. 5⅜ x 8½. 27584-1 Pa. $9.95

PSYCHOLOGY OF MUSIC, Carl E. Seashore. Classic work discusses music as a medium from psychological viewpoint. Clear treatment of physical acoustics, auditory apparatus, sound perception, development of musical skills, nature of musical feeling, host of other topics. 88 figures. 408pp. 5⅜ x 8½. 21851-1 Pa. $11.95

THE PHILOSOPHY OF HISTORY, Georg W. Hegel. Great classic of Western thought develops concept that history is not chance but rational process, the evolution of freedom. 457pp. 5⅜ x 8½. 20112-0 Pa. $9.95

THE BOOK OF TEA, Kakuzo Okakura. Minor classic of the Orient: entertaining, charming explanation, interpretation of traditional Japanese culture in terms of tea ceremony. 94pp. 5⅜ x 8½. 20070-1 Pa. $3.95

LIFE IN ANCIENT EGYPT, Adolf Erman. Fullest, most thorough, detailed older account with much not in more recent books, domestic life, religion, magic, medicine, commerce, much more. Many illustrations reproduce tomb paintings, carvings, hieroglyphs, etc. 597pp. 5⅜ x 8½. 22632-8 Pa. $12.95

SUNDIALS, Their Theory and Construction, Albert Waugh. Far and away the best, most thorough coverage of ideas, mathematics concerned, types, construction, adjusting anywhere. Simple, nontechnical treatment allows even children to build several of these dials. Over 100 illustrations. 230pp. 5⅜ x 8½. 22947-5 Pa. $8.95

THEORETICAL HYDRODYNAMICS, L. M. Milne-Thomson. Classic exposition of the mathematical theory of fluid motion, applicable to both hydrodynamics and aerodynamics. Over 600 exercises. 768pp. 6⅛ x 9¼. 68970-0 Pa. $20.95

SONGS OF EXPERIENCE: Facsimile Reproduction with 26 Plates in Full Color, William Blake. 26 full-color plates from a rare 1826 edition. Includes "The Tyger," "London," "Holy Thursday," and other poems. Printed text of poems. 48pp. 5¼ x 7. 24636-1 Pa. $4.95

OLD-TIME VIGNETTES IN FULL COLOR, Carol Belanger Grafton (ed.). Over 390 charming, often sentimental illustrations, selected from archives of Victorian graphics—pretty women posing, children playing, food, flowers, kittens and puppies, smiling cherubs, birds and butterflies, much more. All copyright-free. 48pp. 9¼ x 12¼. 27269-9 Pa. $7.95

PERSPECTIVE FOR ARTISTS, Rex Vicat Cole. Depth, perspective of sky and sea, shadows, much more, not usually covered. 391 diagrams, 81 reproductions of drawings and paintings. 279pp. 5⅜ x 8½.   22487-2 Pa. $7.95

DRAWING THE LIVING FIGURE, Joseph Sheppard. Innovative approach to artistic anatomy focuses on specifics of surface anatomy, rather than muscles and bones. Over 170 drawings of live models in front, back and side views, and in widely varying poses. Accompanying diagrams. 177 illustrations. Introduction. Index. 144pp. 8⅜ x11¼.   26723-7 Pa. $8.95

GOTHIC AND OLD ENGLISH ALPHABETS: 100 Complete Fonts, Dan X. Solo. Add power, elegance to posters, signs, other graphics with 100 stunning copyright-free alphabets: Blackstone, Dolbey, Germania, 97 more—including many lower-case, numerals, punctuation marks. 104pp. 8⅛ x 11.   24695-7 Pa. $8.95

HOW TO DO BEADWORK, Mary White. Fundamental book on craft from simple projects to five-bead chains and woven works. 106 illustrations. 142pp. 5⅜ x 8.   20697-1 Pa. $5.95

THE BOOK OF WOOD CARVING, Charles Marshall Sayers. Finest book for beginners discusses fundamentals and offers 34 designs. "Absolutely first rate . . . well thought out and well executed."–E. J. Tangerman. 118pp. 7¾ x 10⅝.
                                                             23654-4 Pa. $7.95

ILLUSTRATED CATALOG OF CIVIL WAR MILITARY GOODS: Union Army Weapons, Insignia, Uniform Accessories, and Other Equipment, Schuyler, Hartley, and Graham. Rare, profusely illustrated 1846 catalog includes Union Army uniform and dress regulations, arms and ammunition, coats, insignia, flags, swords, rifles, etc. 226 illustrations. 160pp. 9 x 12.   24939-5 Pa. $10.95

WOMEN'S FASHIONS OF THE EARLY 1900s: An Unabridged Republication of "New York Fashions, 1909," National Cloak & Suit Co. Rare catalog of mail-order fashions documents women's and children's clothing styles shortly after the turn of the century. Captions offer full descriptions, prices. Invaluable resource for fashion, costume historians. Approximately 725 illustrations. 128pp. 8⅜ x 11¼.
                                                             27276-1 Pa. $11.95

THE 1912 AND 1915 GUSTAV STICKLEY FURNITURE CATALOGS, Gustav Stickley. With over 200 detailed illustrations and descriptions, these two catalogs are essential reading and reference materials and identification guides for Stickley furniture. Captions cite materials, dimensions and prices. 112pp. 6½ x 9¼.
                                                             26676-1 Pa. $9.95

EARLY AMERICAN LOCOMOTIVES, John H. White, Jr. Finest locomotive engravings from early 19th century: historical (1804–74), main-line (after 1870), special, foreign, etc. 147 plates. 142pp. 11⅜ x 8¼.   22772-3 Pa. $10.95

THE TALL SHIPS OF TODAY IN PHOTOGRAPHS, Frank O. Braynard. Lavishly illustrated tribute to nearly 100 majestic contemporary sailing vessels: Amerigo Vespucci, Clearwater, Constitution, Eagle, Mayflower, Sea Cloud, Victory, many more. Authoritative captions provide statistics, background on each ship. 190 black-and-white photographs and illustrations. Introduction. 128pp. 8⅞ x 11¾.
                                                             27163-3 Pa. $14.95

LITTLE BOOK OF EARLY AMERICAN CRAFTS AND TRADES, Peter Stockham (ed.). 1807 children's book explains crafts and trades: baker, hatter, cooper, potter, and many others. 23 copperplate illustrations. 140pp. 4⅝ x 6.
23336-7 Pa. $4.95

VICTORIAN FASHIONS AND COSTUMES FROM HARPER'S BAZAR, 1867–1898, Stella Blum (ed.). Day costumes, evening wear, sports clothes, shoes, hats, other accessories in over 1,000 detailed engravings. 320pp. 9⅜ x 12¼.
22990-4 Pa. $15.95

GUSTAV STICKLEY, THE CRAFTSMAN, Mary Ann Smith. Superb study surveys broad scope of Stickley's achievement, especially in architecture. Design philosophy, rise and fall of the Craftsman empire, descriptions and floor plans for many Craftsman houses, more. 86 black-and-white halftones. 31 line illustrations. Introduction 208pp. 6½ x 9¼. 27210-9 Pa. $9.95

THE LONG ISLAND RAIL ROAD IN EARLY PHOTOGRAPHS, Ron Ziel. Over 220 rare photos, informative text document origin (1844) and development of rail service on Long Island. Vintage views of early trains, locomotives, stations, passengers, crews, much more. Captions. 8⅞ x 11¾. 26301-0 Pa. $13.95

VOYAGE OF THE LIBERDADE, Joshua Slocum. Great 19th-century mariner's thrilling, first-hand account of the wreck of his ship off South America, the 35-foot boat he built from the wreckage, and its remarkable voyage home. 128pp. 5⅜ x 8½.
40022-0 Pa. $4.95

TEN BOOKS ON ARCHITECTURE, Vitruvius. The most important book ever written on architecture. Early Roman aesthetics, technology, classical orders, site selection, all other aspects. Morgan translation. 331pp. 5⅜ x 8½. 20645-9 Pa. $8.95

THE HUMAN FIGURE IN MOTION, Eadweard Muybridge. More than 4,500 stopped-action photos, in action series, showing undraped men, women, children jumping, lying down, throwing, sitting, wrestling, carrying, etc. 390pp. 7⅞ x 10⅝.
20204-6 Clothbd. $27.95

TREES OF THE EASTERN AND CENTRAL UNITED STATES AND CANADA, William M. Harlow. Best one-volume guide to 140 trees. Full descriptions, woodlore, range, etc. Over 600 illustrations. Handy size. 288pp. 4½ x 6⅜.
20395-6 Pa. $6.95

SONGS OF WESTERN BIRDS, Dr. Donald J. Borror. Complete song and call repertoire of 60 western species, including flycatchers, juncoes, cactus wrens, many more–includes fully illustrated booklet. Cassette and manual 99913-0 $8.95

GROWING AND USING HERBS AND SPICES, Milo Miloradovich. Versatile handbook provides all the information needed for cultivation and use of all the herbs and spices available in North America. 4 illustrations. Index. Glossary. 236pp. 5⅜ x 8½.
25058-X Pa. $7.95

BIG BOOK OF MAZES AND LABYRINTHS, Walter Shepherd. 50 mazes and labyrinths in all–classical, solid, ripple, and more–in one great volume. Perfect inexpensive puzzler for clever youngsters. Full solutions. 112pp. 8⅛ x 11.
22951-3 Pa. $5.95

PIANO TUNING, J. Cree Fischer. Clearest, best book for beginner, amateur. Simple repairs, raising dropped notes, tuning by easy method of flattened fifths. No previous skills needed. 4 illustrations. 201pp. 5⅜ x 8½. 23267-0 Pa. $6.95

HINTS TO SINGERS, Lillian Nordica. Selecting the right teacher, developing confidence, overcoming stage fright, and many other important skills receive thoughtful discussion in this indispensible guide, written by a world-famous diva of four decades' experience. 96pp. 5⅜ x 8½. 40094-8 Pa. $4.95

THE COMPLETE NONSENSE OF EDWARD LEAR, Edward Lear. All nonsense limericks, zany alphabets, Owl and Pussycat, songs, nonsense botany, etc., illustrated by Lear. Total of 320pp. 5⅜ x 8½. (USO) 20167-8 Pa. $7.95

VICTORIAN PARLOUR POETRY: An Annotated Anthology, Michael R. Turner. 117 gems by Longfellow, Tennyson, Browning, many lesser-known poets. "The Village Blacksmith," "Curfew Must Not Ring Tonight," "Only a Baby Small," dozens more, often difficult to find elsewhere. Index of poets, titles, first lines. xxiii + 325pp. 5⅜ x 8¼. 27044-0 Pa. $8.95

DUBLINERS, James Joyce. Fifteen stories offer vivid, tightly focused observations of the lives of Dublin's poorer classes. At least one, "The Dead," is considered a masterpiece. Reprinted complete and unabridged from standard edition. 160pp. 5$\frac{3}{16}$ x 8¼. 26870-5 Pa. $1.00

GREAT WEIRD TALES: 14 Stories by Lovecraft, Blackwood, Machen and Others, S. T. Joshi (ed.). 14 spellbinding tales, including "The Sin Eater," by Fiona McLeod, "The Eye Above the Mantel," by Frank Belknap Long, as well as renowned works by R. H. Barlow, Lord Dunsany, Arthur Machen, W. C. Morrow and eight other masters of the genre. 256pp. 5⅜ x 8½. (USO) 40436-6 Pa. $8.95

THE BOOK OF THE SACRED MAGIC OF ABRAMELIN THE MAGE, translated by S. MacGregor Mathers. Medieval manuscript of ceremonial magic. Basic document in Aleister Crowley, Golden Dawn groups. 268pp. 5⅜ x 8½. 23211-5 Pa. $9.95

NEW RUSSIAN-ENGLISH AND ENGLISH-RUSSIAN DICTIONARY, M. A. O'Brien. This is a remarkably handy Russian dictionary, containing a surprising amount of information, including over 70,000 entries. 366pp. 4½ x 6⅛. 20208-9 Pa. $10.95

HISTORIC HOMES OF THE AMERICAN PRESIDENTS, Second, Revised Edition, Irvin Haas. A traveler's guide to American Presidential homes, most open to the public, depicting and describing homes occupied by every American President from George Washington to George Bush. With visiting hours, admission charges, travel routes. 175 photographs. Index. 160pp. 8¼ x 11. 26751-2 Pa. $11.95

NEW YORK IN THE FORTIES, Andreas Feininger. 162 brilliant photographs by the well-known photographer, formerly with *Life* magazine. Commuters, shoppers, Times Square at night, much else from city at its peak. Captions by John von Hartz. 181pp. 9¼ x 10¾. 23585-8 Pa. $13.95

INDIAN SIGN LANGUAGE, William Tomkins. Over 525 signs developed by Sioux and other tribes. Written instructions and diagrams. Also 290 pictographs. 111pp. 6⅛ x 9¼. 22029-X Pa. $3.95

ANATOMY: A Complete Guide for Artists, Joseph Sheppard. A master of figure drawing shows artists how to render human anatomy convincingly. Over 460 illustrations. 224pp. 8⅜ x 11¼. 27279-6 Pa. $11.95

MEDIEVAL CALLIGRAPHY: Its History and Technique, Marc Drogin. Spirited history, comprehensive instruction manual covers 13 styles (ca. 4th century thru 15th). Excellent photographs; directions for duplicating medieval techniques with modern tools. 224pp. 8⅜ x 11¼. 26142-5 Pa. $12.95

DRIED FLOWERS: How to Prepare Them, Sarah Whitlock and Martha Rankin. Complete instructions on how to use silica gel, meal and borax, perlite aggregate, sand and borax, glycerine and water to create attractive permanent flower arrangements. 12 illustrations. 32pp. 5⅜ x 8½. 21802-3 Pa. $1.00

EASY-TO-MAKE BIRD FEEDERS FOR WOODWORKERS, Scott D. Campbell. Detailed, simple-to-use guide for designing, constructing, caring for and using feeders. Text, illustrations for 12 classic and contemporary designs. 96pp. 5⅜ x 8½. 25847-5 Pa. $3.95

SCOTTISH WONDER TALES FROM MYTH AND LEGEND, Donald A. Mackenzie. 16 lively tales tell of giants rumbling down mountainsides, of a magic wand that turns stone pillars into warriors, of gods and goddesses, evil hags, powerful forces and more. 240pp. 5⅜ x 8½. 29677-6 Pa. $6.95

THE HISTORY OF UNDERCLOTHES, C. Willett Cunnington and Phyllis Cunnington. Fascinating, well-documented survey covering six centuries of English undergarments, enhanced with over 100 illustrations: 12th-century laced-up bodice, footed long drawers (1795), 19th-century bustles, 19th-century corsets for men, Victorian "bust improvers," much more. 272pp. 5⅜ x 8¼. 27124-2 Pa. $9.95

ARTS AND CRAFTS FURNITURE: The Complete Brooks Catalog of 1912, Brooks Manufacturing Co. Photos and detailed descriptions of more than 150 now very collectible furniture designs from the Arts and Crafts movement depict davenports, settees, buffets, desks, tables, chairs, bedsteads, dressers and more, all built of solid, quarter-sawed oak. Invaluable for students and enthusiasts of antiques, Americana and the decorative arts. 80pp. 6½ x 9¼. 27471-3 Pa. $8.95

WILBUR AND ORVILLE: A Biography of the Wright Brothers, Fred Howard. Definitive, crisply written study tells the full story of the brothers' lives and work. A vividly written biography, unparalleled in scope and color, that also captures the spirit of an extraordinary era. 560pp. 6⅛ x 9¼. 40297-5 Pa. $17.95

THE ARTS OF THE SAILOR: Knotting, Splicing and Ropework, Hervey Garrett Smith. Indispensable shipboard reference covers tools, basic knots and useful hitches; handsewing and canvas work, more. Over 100 illustrations. Delightful reading for sea lovers. 256pp. 5⅜ x 8½. 26440-8 Pa. $8.95

FRANK LLOYD WRIGHT'S FALLINGWATER: The House and Its History, Second, Revised Edition, Donald Hoffmann. A total revision–both in text and illustrations–of the standard document on Fallingwater, the boldest, most personal architectural statement of Wright's mature years, updated with valuable new material from the recently opened Frank Lloyd Wright Archives. "Fascinating"–*The New York Times.* 116 illustrations. 128pp. 9¼ x 10¾. 27430-6 Pa. $12.95

CATALOG OF DOVER BOOKS

PHOTOGRAPHIC SKETCHBOOK OF THE CIVIL WAR, Alexander Gardner. 100 photos taken on field during the Civil War. Famous shots of Manassas Harper's Ferry, Lincoln, Richmond, slave pens, etc. 244pp. 10⅛ x 8¼.   22731-6 Pa. $10.95

FIVE ACRES AND INDEPENDENCE, Maurice G. Kains. Great back-to-the-land classic explains basics of self-sufficient farming. The one book to get. 95 illustrations. 397pp. 5⅜ x 8½.   20974-1 Pa. $7.95

SONGS OF EASTERN BIRDS, Dr. Donald J. Borror. Songs and calls of 60 species most common to eastern U.S.: warblers, woodpeckers, flycatchers, thrushes, larks, many more in high-quality recording.   Cassette and manual 99912-2 $9.95

A MODERN HERBAL, Margaret Grieve. Much the fullest, most exact, most useful compilation of herbal material. Gigantic alphabetical encyclopedia, from aconite to zedoary, gives botanical information, medical properties, folklore, economic uses, much else. Indispensable to serious reader. 161 illustrations. 888pp. 6½ x 9¼. 2-vol. set. (USO)   Vol. I: 22798-7 Pa. $9.95

    Vol. II: 22799-5 Pa. $9.95

HIDDEN TREASURE MAZE BOOK, Dave Phillips. Solve 34 challenging mazes accompanied by heroic tales of adventure. Evil dragons, people-eating plants, blood-thirsty giants, many more dangerous adversaries lurk at every twist and turn. 34 mazes, stories, solutions. 48pp. 8¼ x 11.   24566-7 Pa. $2.95

LETTERS OF W. A. MOZART, Wolfgang A. Mozart. Remarkable letters show bawdy wit, humor, imagination, musical insights, contemporary musical world; includes some letters from Leopold Mozart. 276pp. 5⅜ x 8½.   22859-2 Pa. $7.95

BASIC PRINCIPLES OF CLASSICAL BALLET, Agrippina Vaganova. Great Russian theoretician, teacher explains methods for teaching classical ballet. 118 illustrations. 175pp. 5⅜ x 8½.   22036-2 Pa. $5.95

THE JUMPING FROG, Mark Twain. Revenge edition. The original story of The Celebrated Jumping Frog of Calaveras County, a hapless French translation, and Twain's hilarious "retranslation" from the French. 12 illustrations. 66pp. 5⅜ x 8½.   22686-7 Pa. $3.95

BEST REMEMBERED POEMS, Martin Gardner (ed.). The 126 poems in this superb collection of 19th- and 20th-century British and American verse range from Shelley's "To a Skylark" to the impassioned "Renascence" of Edna St. Vincent Millay and to Edward Lear's whimsical "The Owl and the Pussycat." 224pp. 5⅜ x 8½.   27165-X Pa. $5.95

COMPLETE SONNETS, William Shakespeare. Over 150 exquisite poems deal with love, friendship, the tyranny of time, beauty's evanescence, death and other themes in language of remarkable power, precision and beauty. Glossary of archaic terms. 80pp. 5³⁄₁₆ x 8¼.   26686-9 Pa. $1.00

BODIES IN A BOOKSHOP, R. T. Campbell. Challenging mystery of blackmail and murder with ingenious plot and superbly drawn characters. In the best tradition of British suspense fiction. 192pp. 5⅜ x 8½.   24720-1 Pa. $6.95

THE WIT AND HUMOR OF OSCAR WILDE, Alvin Redman (ed.). More than 1,000 ripostes, paradoxes, wisecracks: Work is the curse of the drinking classes; I can resist everything except temptation; etc. 258pp. 5⅜ x 8½. 20602-5 Pa. $6.95

SHAKESPEARE LEXICON AND QUOTATION DICTIONARY, Alexander Schmidt. Full definitions, locations, shades of meaning in every word in plays and poems. More than 50,000 exact quotations. 1,485pp. 6½ x 9¼. 2-vol. set.
Vol. 1: 22726-X Pa. $17.95
Vol. 2: 22727-8 Pa. $17.95

SELECTED POEMS, Emily Dickinson. Over 100 best-known, best-loved poems by one of America's foremost poets, reprinted from authoritative early editions. No comparable edition at this price. Index of first lines. 64pp. 5³⁄₁₆ x 8¼.
26466-1 Pa. $1.00

THE INSIDIOUS DR. FU-MANCHU, Sax Rohmer. The first of the popular mystery series introduces a pair of English detectives to their archnemesis, the diabolical Dr. Fu-Manchu. Flavorful atmosphere, fast-paced action, and colorful characters enliven this classic of the genre. 208pp. 5³⁄₁₆ x 8¼. 29898-1 Pa. $2.00

THE MALLEUS MALEFICARUM OF KRAMER AND SPRENGER, translated by Montague Summers. Full text of most important witchhunter's "bible," used by both Catholics and Protestants. 278pp. 6⅝ x 10. 22802-9 Pa. $12.95

SPANISH STORIES/CUENTOS ESPAÑOLES: A Dual-Language Book, Angel Flores (ed.). Unique format offers 13 great stories in Spanish by Cervantes, Borges, others. Faithful English translations on facing pages. 352pp. 5⅜ x 8½.
25399-6 Pa. $8.95

GARDEN CITY, LONG ISLAND, IN EARLY PHOTOGRAPHS, 1869–1919, Mildred H. Smith. Handsome treasury of 118 vintage pictures, accompanied by carefully researched captions, document the Garden City Hotel fire (1899), the Vanderbilt Cup Race (1908), the first airmail flight departing from the Nassau Boulevard Aerodrome (1911), and much more. 96pp. 8⅞ x 11¾. 40669-5 Pa. $12.95

OLD QUEENS, N.Y., IN EARLY PHOTOGRAPHS, Vincent F. Seyfried and William Asadorian. Over 160 rare photographs of Maspeth, Jamaica, Jackson Heights, and other areas. Vintage views of DeWitt Clinton mansion, 1939 World's Fair and more. Captions. 192pp. 8⅞ x 11. 26358-4 Pa. $12.95

CAPTURED BY THE INDIANS: 15 Firsthand Accounts, 1750-1870, Frederick Drimmer. Astounding true historical accounts of grisly torture, bloody conflicts, relentless pursuits, miraculous escapes and more, by people who lived to tell the tale. 384pp. 5⅜ x 8½. 24901-8 Pa. $8.95

THE WORLD'S GREAT SPEECHES (Fourth Enlarged Edition), Lewis Copeland, Lawrence W. Lamm, and Stephen J. McKenna. Nearly 300 speeches provide public speakers with a wealth of updated quotes and inspiration–from Pericles' funeral oration and William Jennings Bryan's "Cross of Gold Speech" to Malcolm X's powerful words on the Black Revolution and Earl of Spenser's tribute to his sister, Diana, Princess of Wales. 944pp. 5⅜ x 8⅜. 40903-1 Pa. $15.95

THE BOOK OF THE SWORD, Sir Richard F. Burton. Great Victorian scholar/adventurer's eloquent, erudite history of the "queen of weapons"–from prehistory to early Roman Empire. Evolution and development of early swords, variations (sabre, broadsword, cutlass, scimitar, etc.), much more. 336pp. 6⅛ x 9¼.
25434-8 Pa. $9.95

AUTOBIOGRAPHY: The Story of My Experiments with Truth, Mohandas K. Gandhi. Boyhood, legal studies, purification, the growth of the Satyagraha (nonviolent protest) movement. Critical, inspiring work of the man responsible for the freedom of India. 480pp. 5⅜ x 8½. (USO) 24593-4 Pa. $8.95

CELTIC MYTHS AND LEGENDS, T. W. Rolleston. Masterful retelling of Irish and Welsh stories and tales. Cuchulain, King Arthur, Deirdre, the Grail, many more. First paperback edition. 58 full-page illustrations. 512pp. 5⅜ x 8½. 26507-2 Pa. $9.95

THE PRINCIPLES OF PSYCHOLOGY, William James. Famous long course complete, unabridged. Stream of thought, time perception, memory, experimental methods; great work decades ahead of its time. 94 figures. 1,391pp. 5⅜ x 8½. 2-vol. set.
Vol. I: 20381-6 Pa. $13.95
Vol. II: 20382-4 Pa. $14.95

THE WORLD AS WILL AND REPRESENTATION, Arthur Schopenhauer. Definitive English translation of Schopenhauer's life work, correcting more than 1,000 errors, omissions in earlier translations. Translated by E. F. J. Payne. Total of 1,269pp. 5⅜ x 8½. 2-vol. set. Vol. 1: 21761-2 Pa. $12.95
Vol. 2: 21762-0 Pa. $12.95

MAGIC AND MYSTERY IN TIBET, Madame Alexandra David-Neel. Experiences among lamas, magicians, sages, sorcerers, Bonpa wizards. A true psychic discovery. 32 illustrations. 321pp. 5⅜ x 8½. (USO) 22682-4 Pa. $9.95

THE EGYPTIAN BOOK OF THE DEAD, E. A. Wallis Budge. Complete reproduction of Ani's papyrus, finest ever found. Full hieroglyphic text, interlinear transliteration, word-for-word translation, smooth translation. 533pp. 6½ x 9¼.
21866-X Pa. $11.95

MATHEMATICS FOR THE NONMATHEMATICIAN, Morris Kline. Detailed, college-level treatment of mathematics in cultural and historical context, with numerous exercises. Recommended Reading Lists. Tables. Numerous figures. 641pp. 5⅜ x 8½.
24823-2 Pa. $11.95

PROBABILISTIC METHODS IN THE THEORY OF STRUCTURES, Isaac Elishakoff. Well-written introduction covers the elements of the theory of probability from two or more random variables, the reliability of such multivariable structures, the theory of random function, Monte Carlo methods of treating problems incapable of exact solution, and more. Examples. 502pp. 5³/₈ x 8¹/₂. 40691-1 Pa. $16.95

THE RIME OF THE ANCIENT MARINER, Gustave Doré, S. T. Coleridge. Doré's finest work; 34 plates capture moods, subtleties of poem. Flawless full-size reproductions printed on facing pages with authoritative text of poem. "Beautiful. Simply beautiful."–*Publisher's Weekly.* 77pp. 9¼ x 12. 22305-1 Pa. $7.95

NORTH AMERICAN INDIAN DESIGNS FOR ARTISTS AND CRAFTSPEOPLE, Eva Wilson. Over 360 authentic copyright-free designs adapted from Navajo blankets, Hopi pottery, Sioux buffalo hides, more. Geometrics, symbolic figures, plant and animal motifs, etc. 128pp. 8⅜ x 11. (EUK) 25341-4 Pa. $8.95

SCULPTURE: Principles and Practice, Louis Slobodkin. Step-by-step approach to clay, plaster, metals, stone; classical and modern. 253 drawings, photos. 255pp. 8⅛ x 11.
22960-2 Pa. $11.95

THE INFLUENCE OF SEA POWER UPON HISTORY, 1660–1783, A. T. Mahan. Influential classic of naval history and tactics still used as text in war colleges. First paperback edition. 4 maps. 24 battle plans. 640pp. 5⅜ x 8½. 25509-3 Pa. $14.95

THE STORY OF THE TITANIC AS TOLD BY ITS SURVIVORS, Jack Winocour (ed.). What it was really like. Panic, despair, shocking inefficiency, and a little heroism. More thrilling than any fictional account. 26 illustrations. 320pp. 5⅜ x 8½. 20610-6 Pa. $8.95

FAIRY AND FOLK TALES OF THE IRISH PEASANTRY, William Butler Yeats (ed.). Treasury of 64 tales from the twilight world of Celtic myth and legend: "The Soul Cages," "The Kildare Pooka," "King O'Toole and his Goose," many more. Introduction and Notes by W. B. Yeats. 352pp. 5⅜ x 8½. 26941-8 Pa. $8.95

BUDDHIST MAHAYANA TEXTS, E. B. Cowell and Others (eds.). Superb, accurate translations of basic documents in Mahayana Buddhism, highly important in history of religions. The Buddha-karita of Asvaghosha, Larger Sukhavativyuha, more. 448pp. 5⅜ x 8½. 25552-2 Pa. $12.95

ONE TWO THREE . . . INFINITY: Facts and Speculations of Science, George Gamow. Great physicist's fascinating, readable overview of contemporary science: number theory, relativity, fourth dimension, entropy, genes, atomic structure, much more. 128 illustrations. Index. 352pp. 5⅜ x 8½. 25664-2 Pa. $8.95

EXPERIMENTATION AND MEASUREMENT, W. J. Youden. Introductory manual explains laws of measurement in simple terms and offers tips for achieving accuracy and minimizing errors. Mathematics of measurement, use of instruments, experimenting with machines. 1994 edition. Foreword. Preface. Introduction. Epilogue. Selected Readings. Glossary. Index. Tables and figures. 128pp. 5³/₈ x 8¹/₂. 40451-X Pa. $6.95

DALÍ ON MODERN ART: The Cuckolds of Antiquated Modern Art, Salvador Dalí. Influential painter skewers modern art and its practitioners. Outrageous evaluations of Picasso, Cézanne, Turner, more. 15 renderings of paintings discussed. 44 calligraphic decorations by Dalí. 96pp. 5⅜ x 8½. (USO) 29220-7 Pa. $5.95

ANTIQUE PLAYING CARDS: A Pictorial History, Henry René D'Allemagne. Over 900 elaborate, decorative images from rare playing cards (14th–20th centuries): Bacchus, death, dancing dogs, hunting scenes, royal coats of arms, players cheating, much more. 96pp. 9¼ x 12¼. 29265-7 Pa. $12.95

MAKING FURNITURE MASTERPIECES: 30 Projects with Measured Drawings, Franklin H. Gottshall. Step-by-step instructions, illustrations for constructing handsome, useful pieces, among them a Sheraton desk, Chippendale chair, Spanish desk, Queen Anne table and a William and Mary dressing mirror. 224pp. 8⅛ x 11¼. 29338-6 Pa. $13.95

THE FOSSIL BOOK: A Record of Prehistoric Life, Patricia V. Rich et al. Profusely illustrated definitive guide covers everything from single-celled organisms and dinosaurs to birds and mammals and the interplay between climate and man. Over 1,500 illustrations. 760pp. 7½ x 10⅛. 29371-8 Pa. $29.95

Prices subject to change without notice.

Available at your book dealer or write for free catalog to Dept. GI, Dover Publications, Inc., 31 East 2nd St., Mineola, N.Y. 11501. Dover publishes more than 500 books each year on science, elementary and advanced mathematics, biology, music, art, literary history, social sciences and other areas.